WILD ANIMAL SKINS IN VICTORIAN BRITAIN

To my lifelong friend Susan Mitchell (née Poole).

Our friendship was not only founded before we were born by a community of blood, but is in itself near as old as my life. It began with our early ages, and, like a history, has been continued to the present time. Although we may not be old in the world we are old to each other, having so long been intimates. We are now widely separated, a great sea and continent intervening; but memory, like care, mounts into iron ships and rides post behind the horseman. Neither time nor space nor enmity can conquer old affection; and as I dedicate these sketches, it is not to you only, but to all in the old country, that I send the greeting of my heart.

Robert Louis Stevenson, 1879
(Stevenson's dedication of *Travels with a Donkey*
to Robert Alan Mowbray Stevenson)

Wild Animal Skins in Victorian Britain

Zoos, Collections, Portraits, and Maps

ANN C. COLLEY
State University College of New York at Buffalo, USA

Routledge
Taylor & Francis Group
LONDON AND NEW YORK

First published in paperback 2024

First published 2014 by Ashgate Publishing

Published 2016 by Routledge
4 Park Square, Milton Park, Abingdon, Oxon OX14 4RN

and by Routledge
605 Third Avenue, New York, NY 10158

Routledge is an imprint of the Taylor & Francis Group, an informa business

British Library Cataloguing in Publication Data
A catalogue record for this book is available from the British Library.

The Library of Congress has cataloged the printed edition as follows:
Colley, Ann C.
Wild animal skins in Victorian Britain : zoos, collections, portraits, and maps / by Ann C. Colley.

pages cm
Includes bibliographical references and index.
ISBN 978-1-4724-2778-6 (hardcover)
1. Zoological specimens--Collection and preservation--Great Britain--History--Victoria, 1837-1901. 2. Hides and skins--Collection and preservation--Great Britain--History--Victoria, 1837-1901. 3. Animals in art. 4. Art, Victorian. I. Title.
QL67.C65 2014
590.75'2--dc23

2014026179

ISBN: 978-1-4724-2778-6 (hbk)
ISBN: 978-1-03-292709-1 (pbk)
ISBN: 978-1-315-54717-6 (ebk)

DOI: 10.4324/9781315547176

Contents

List of Figures *vii*
List of Plates *ix*
Acknowledgments *xi*

Introduction 1

Preamble: Theorizing about Skin 9

1 Industry, Empire, Portraiture, and Skin at the Belle Vue Zoo, Manchester 17

2 A Skin Disorder 49

3 Stuff and Nonsense: Skin and Victorian Animal Portraiture 87

4 Touch: Reaching through the Bars 121

5 Wild Skins and Mapping the Victorian Landscape 149

Bibliography *183*
Index *195*

List of Figures

I.1	"Irene and School Friends"	2
I.2	"Furs"	3
1.1	*Manchester from Belle Vue*	19
1.2	"Matabele War"	25
1.3	"Delhi: Storming the Kashmir Gate"	25
1.4	"Mons 1914–1918"	26
1.5	"Zoological Gardens, Belle Vue, Manchester."	26
1.6	"Aptenodytes Pennantis, Esq."	36
1.7	"Coming to the Point"	37
1.8	"Pelican Enclosure"	38
1.9	"At the Zoo"	39
1.10	"Elephant Skeleton in Belle Vue Museum"	44
2.1	Cartoon from *Jugend*	54
2.2	"Keeper's Nightmare"	55
2.3	"Troglodytes Gorilla"	58
2.4	"A.D. Bartlett with his first preserved gorilla"	59
2.5	"Escaped Kangaroo at Regent's Park"	61
2.6	"Discomfiture of Old Mr. J—N—S"	66
2.7	"Skinning a Tiger"	68
2.8	"Ah me! My uncle's spirit is in these stones"	81
2.9	"Quiere comprar un Condor"	83
3.1/3.2	From Edward Lear's "The Story of Four Little Children Who Went Round the World"	90
3.3	Wardian Furniture	92
3.4	Plate III from *The Taxidermist's Manual*	99
3.5	"Manikin for Zebra, Completed, Ready for the Skin"	102
3.6	"The Specimen Completed"	103
3.7	"Si Non E Vero Etc."	105
3.8	Drawing of Kiwi, 1839	108
3.9	Lear's "Old Person of Crowle"	114
3.10	Lear's "There was an Old Man of Dumbree"	115
3.11	Lear's "There was an Old Man with an Owl"	115
3.12	Sketch of unidentified parrot by Edward Lear	117

3.13 Sketch of Phos By Edward Lear 118
3.14 Edward Lear's "There was an Old Man on the Border" 119

4.1 "Child extending hand through bars of rhinoceros enclosure" 122
4.2 Beaver Enclosure 123
4.3 "Best of Friends—Hippopotamus and Keeper" 124
4.4 "Moti and its Keeper" 125
4.5 "Keeper Stroking Tapir" 125
4.6 "A Prospecte of Ye Zoological Societye: Its Gardens" 127
4.7 Adventures of the Ojibbeway and Ioway Indians in England, France, and Belgium 131
4.8 "The Polar Bear Rider" 138

5.1 Physical Map of the World from the Discoveries of the Most Eminent Modern Geographers 153
5.2 End paper from Sport in the Highlands of Kashmir 158
5.3 Zoological Map of the World, Shewing the Geographical Distribution of Animals 162
5.4 Zoological Map Showing the Distribution of Animals over the World, and Zoological Map Showing the Distribution of Birds & Reptiles over the World 164
5.5 "This is a glacier, though you wd. not think it. July 13, '68" 168
5.6 Sketch accompanying Sunday, April 10 (1864) entry in Hopkins's *Journals* 169
5.7 Sketch accompanying July 2 (1868) entry in Hopkins's *Journals* 172
5.8 Sketch accompanying July 10 (1868) entry in Hopkins's *Journals* 172
5.9 "Ice on my tadpole basin formed as below" 172
5.10 "Sept 4, '68" from Sketchbook D 174

List of Plates

Plate 1	"Rhinoceros, Zoological Gardens, London"	122
Plate 2	"Camel Ride"	124
Plate 3	Sir Lawrence Alma-Tadema's *In the Tepidarium* (1881)	141
Plate 4	James McNeill Whistler's *Symphony in White, No. 1: The White Girl* (1862)	144

Acknowledgments

Many people and places have made this book possible. I would like to thank Clemency Fisher, Department of Vertebrate Zoology, the World Museum, National Museums Liverpool. Clem's enthusiasm for her subject and her generosity toward me were extraordinary. She made room for me in her overcrowded office, and when she was off for a couple of days arranged for me to continue my research in another space (among staff members who were stuffing spiders!). I am also grateful to Christine E. Jackson, who has transcribed many of Elizabeth Hornby's letters. While in Liverpool my research was also helped by John Edmondson, who arranged for me to visit Knowsley Hall where the then Curator, Tom Boggs, spent a good part of the day showing me the books, watercolors, and prints collected by the 13th Earl of Derby. I am also indebted to Jane Muskett, archivist of Chetham's Library, Manchester. Indeed the entire staff of Chetham's Library was most helpful when I was reading through the Belle Vue Zoo archives. (There I sat in the chairs once occupied by Karl Marx and Friedrich Engels.) Similarly I am grateful to the archivists at the Caird Library, the National Maritime Museum (Greenwich) and especially appreciative of the time I spent with William Edwards, the Curator of the Gordon Museum, Guy's Hospital (London), where he showed me Joseph Towne's wax models depicting diseases of the skin. The staff of the Cambridge University Library and the Maps Room of the British Library also consistently offered me well-informed and professional support.

None of this research would have been possible without the support of Ralph Wahlstrom and Dean Benjamin C. Christy, both of whom arranged for my Title F Research Leave from the State University of New York College at Buffalo. And most especially, this work would have not been possible without the support of Wolfson College, Cambridge University. As a Visiting Fellow for the Lent and Easter Terms (2011), I was given the opportunity to gather extraordinary archival materials. Friends and acquaintances were also essential to my project. Among these people are Dr. Christine L. Corton who kept informing me of references, Joel Huberman, and Kaylene Waite, Head of Computer Graphics at my home institution. Others include my childhood friends Susan and Fred Mitchell, who made it possible for me to stay in the Manchester area, and my cousins, Lyn Brown and Jim Cheetham, who looked after me during my time in Liverpool. I would also like to thank Professor Glenda Norquay of Liverpool John Moores University for inviting me to give a talk about my work in progress. The questions posed by her colleagues helped me to find my way through the plethora of materials I was collecting. And I am thankful for the support of Professor Adrienne Munich and her PhD student Kimberly Cox of the English Department, SUNY at Stony Brook for helping me to develop my ideas, particularly those concerning the sense of touch. As always, I have a deep regard for those who encouraged me and read this book while it was in process. The care which Carrie Tirado Bramen (SUNY Buffalo), Regina Grol (Fellow, Center of Slavic, Eurasian, and East European Studies at UNC, Chapel Hill), and Carolyn W. Korsmeyer (SUNY Buffalo) took with the manuscript is remarkable.

Finally I want to acknowledge the Cambridge University Press for giving me permission to reprint parts of my essay "Portrait, Empire, and Industry at the Belle Vue Zoo, Manchester," which originally appeared in *Victorian Literature and Culture*, 42.2 (2014): 167–86. I also wish to thank the Rt. Hon the Earl of Derby, the British Library Board, the Master and Fellows of Balliol College (Oxford), National Museums Liverpool, the Houghton Library (Harvard University), Chetham's Library (Manchester), Irene Haupt, and the Getty Research Institute, Los Angeles for permission to reprint verbal and visual materials. Every effort has been made to trace or contact all copyright holders. I would be pleased to rectify any omissions brought to my notice at the earliest opportunity.

Introduction

I am looking at a photograph of a friend, taken in the mid-1940s. (She stands second on the left among her school friends.) To compose this picture, her father, who was also the photographer, has lined the children up, and notably in front of them, has placed three mounted wild animals: a wildcat, a monkey (maybe a howler), and a flamingo. (My friend remembers that these stuffed skins were usually housed in the storeroom of the local cinema.) Standing in the foreground, these imposing, almost lifelike, creatures preface a view of the children to create a disturbing portrait—an image that becomes even more striking when one notices the parallels between the specimens' luminescent glass eyes turned toward the camera lens and the children's obedient attention to the photographer and his equipment.

Why, one wonders, are these stuffed wild skins and the children posed together? Perhaps my friend's father thought it would be amusing and realized that the children found these animals intriguing. (She says there was no zoo nearby.) The animals' seemingly incongruous presence is, perhaps, also indebted to his belief that humans and animals share this earth. Whatever the reason, their inclusion is a vivid reminder of an epoch in which displays of taxidermy and treated hides were a more ordinary part of people's lives. Especially in the century before this photograph was snapped, exhibits of mammal and bird skins were not uncommon in public spaces. This mid-1940s image is evidence of the afterlife of a time in the nineteenth century when it would *not* have been considered unusual to display the skins of exotic creatures. As such, the prominence of the wildcat, monkey, and flamingo evokes the Victorian fascination with the hides and feathers of exotic creatures, and, furthermore, recalls people's desire to arrange and put them on view for others to see and study. These stuffed creatures in the photograph are leftovers of a period when skin was even more central than it is now to the public's understanding of place and identity—to a time when skin was not only a basic ingredient of portraiture but also the site of encounter with the exotic world.[1]

This book is about the Victorians' fascination with wild skins. In particular, *Wild Animal Skins in Victorian Britain: Zoos, Collections, Portraits, and Maps*

---

[1] As many commentators remark, we now live in a culture that tends to be either repelled or titillated by exhibits of skins. In this respect, I am reminded of a person I recently met in London who told me that when he was a boy in the 1930s, he dreaded going to the dentist because when he sat back in the dentist's chair, the man's collection of stuffed birds on the surrounding shelves became visible. These specimens seemed to stare at him—particularly the owl that glared down at him with its piercing eyes and the raven that bent its foreboding head in his direction.

Figure I.1 "Irene and School Friends"
Source: Haupt.

considers the collecting, displaying, and portraying of exotic creatures in nineteenth-century Britain. This study is not, as the opening paragraphs might suggest, a work exclusively devoted to taxidermy (as several recent works on animal skins are), but rather is about the importance and meaning attributed to wild skins during the Victorian era. In the 1800s, at a time when Britain controlled more than a fifth of the world's land area, new trade and military routes made it possible to collect and exhibit natural history specimens from faraway places only vaguely known to the British public. The skins that caught people's avid attention were not necessarily the commercial products hung and draped in the "Fur" Hall at the 1851 Great Exhibition.[2]

Figure I.2 "Furs"
Source: Getty Research Institute.

[2] When reading *Reports by the Juries* (a book listing and commenting on the items exhibited at the Great Exhibition of 1851), I was fascinated to learn that "the estimated number [of hides, skins, and furs] imported into Great Britain annually is about 11,000,000 of which 5,000,000 are applicable to furs, and 6,000,000 are adapted for various descriptions of leather" (383). I am grateful to Professor Thad Logan of Rice University for pointing out this publication.

Already familiar to those visitors through the manufacturing of rugs and ornamental items for clothing, these marketable furs and plumes were not as compelling as were the less familiar skins and feathers from areas of the empire barely known to the British populace. It was these exotic skins that intrigued the public and made people eager to catch a glimpse of them. As a consequence, zoos proliferated in Britain's major cities; traveling menageries made their way through provincial towns; wealthy estates amassed private collections; and museums displayed innumerable cabinets and cases of specimens. Moreover, scientific institutions sent out collectors to bring back exotic animals and birds so that anatomists and zoologists might study, label, and systematically classify these creatures. This preoccupation with natural history cut across class lines so that both the landed gentry, such as the extraordinarily wealthy 13th Earl of Derby, as well as the ordinary laborer, avidly assembled and, according to their means, displayed their precious specimens. Obviously when gathering specimens for these collections, it was not always possible to acquire live animals or birds, so people generally agreed that the collecting of skins was a more manageable endeavor. Compared to a struggling and all-too-vulnerable living creature, hides and feathers seemed more durable and portable: it was thought that they could more easily be labeled, displayed, or stored; moreover, they could be reshaped into trophies to create a tangible portrait of empire.

Because of the period's dedication to collecting, studying, and portraying exotic skins and because of the age's dedication to the expanse of empire, this study, from time to time, inevitably finds itself situated within this colonial context and participating in a post-colonial critical mode, but not, I hasten to add, uncritically. There is, instead, an attempt to qualify oft-repeated truths concerning the assertion of imperial authority through exhibits of specimens from distant parts of the empire. This study, for instance, points out that this desired control was more often than not unavailable, or at best frustrated, and that the gathering, arranging, transporting, and labeling of skins from foreign territories, rather than being a means of displaying British rule, were instead emblematic of the messiness of empire.

Wild Animal Skins in Victorian Britain, however, is not entirely beholden to the subject's colonial context. As the "Preamble: Theorizing about Skin" explains, the book also breaks through the barriers of a specific time and place in order to consider more generally the function of skin in portraiture or illustration. In this respect, *Wild Skins* not only explores collections and depictions of exotic animal skins but also discusses the ways in which skin represents identity. This study recognizes that, for the Victorians, skin was a worn identity and was, therefore, essential to determining a species' distinctiveness. Skin was also seen as a protective, yet vulnerable, envelope, which metonymically contained and represented a being's entirety. Moreover, *Wild Animal Skins* recognizes the cognitive function of skin and the ways in which skin proffers access not only to another being but also to foreign territories.

Because so many areas of Victorian culture are involved in this subject, the book calls upon several points of view: literary texts written by such writers as Charles Dickens, Elizabeth Gaskell, Thomas Hardy, Rudyard Kipling, Edward Lear, and Gerard Manley Hopkins; nineteenth-century zoo archives; scientific treatises; theories of portraiture; theater history; unpublished diaries and letters between collectors and their agents; adventure fiction; Victorian travel narratives; cartoons and articles from nineteenth-century journals; portraits by such painters as Sir Lawrence Alma-Tadema and James McNeill Whistler; natural history illustrations; nonsense drawings and verse; critical studies of skin and touch; taxidermy manuals; as well as Victorian maps made by adventurers and scientists.

Wild Animal Skins in Victorian Britain opens by taking the reader into the Belle Vue Zoo, Manchester, a place where the general public would have had a chance to gaze at living and stuffed skins from the reach of empire. The first chapter, "Industry, Empire, Portraiture, and Skin at the Belle Vue Zoo, Manchester," looks at this provincial zoo, founded in 1836, which was dedicated to promoting and glorifying Britain's industrial and colonial achievements. To celebrate these, Belle Vue assembled a series of "portrait galleries." These galleries displayed pictures, among which were those that depended not only on the more conventional forms of portraiture (found in places such as the National Portrait Gallery) but also on the medium and display of wild skins. At Belle Vue, skin was more than just a tangible souvenir of conquest; it was also a compelling ingredient of portraiture and, therefore, identity.

Chapter 2, "A Skin Disorder," focuses on the collecting of animal and bird skins from the reach of empire. This chapter explores the muddles, disappointments, and disasters inherent in Victorian natural history collecting; it corrects the received belief that when amassing and arranging these spoils, the British exercised control over foreign territories. Accounts written by Victorian travelers, hunters, and agents demonstrate that the opposite was true: mastery was rarely realized. To illustrate further the attendant frustrations and disorder accompanying the collecting of specimens, an Afterword "In the Field," examines the unpublished correspondence and notebooks of two avid nineteenth-century collectors: the 13th Earl of Derby, and his 22-year-old niece, Elizabeth Hornby, who collected and stuffed wild skins for her own modest display as well as for her uncle's more ambitious menagerie and museum while she resided in South America from 1847 to 1850.

Chapter 3, "Stuff and Nonsense: Skin and Victorian Animal Portraiture," uses the occasion of Edward Lear's relationship with the 13th Earl of Derby (Lear was hired by him to illustrate the exotic animals kept on his 45,000-acre estate) as well as one of Lear's nonsense verses ("The Story of the Four Little Children Who Went Round the World"), which was based upon his experiences while working for Lord Derby, to consider the centrality of skin in Victorian animal portraiture and taxidermy. The chapter offers an occasion to think about the function as well as the meaning of skin in taxidermy and to focus on the depiction of skin in natural history portraits. The discussion of this topic leads to the suggestion that Lear, through his nonsense verses, as well as in his natural history illustrations, rebelled against the almost exclusive attention to the skins or surfaces of the portrayed animals. He preferred to get "under the skin" and proffer a glimpse of a creature's subjectivity. As if releasing his subjects from the conventions of nineteenth-century natural history illustration and portraiture, Lear challenged the colonial's or the collector's commanding gaze and liberated his subjects from the prerogatives of classification, ownership, and commodity.

Chapter 4, "Touch: Reaching through the Bars," is about the desire to touch wild skins. This chapter opens with accounts of the propensity, on the part of the Victorian public, to extend their hands through the bars of cages in zoos so as to be able to "caress" the enclosed creature's fur, to brush against a bird's feathers, or to feel the lick of the wild animal's tongue. This desire to link skin to skin gave visitors direct access to the exotic other (to its texture, temperature, movements, and pressure). In addition, it put members of the public in touch with their own wildness (not only when they rubbed up against a bear's hide but also when they stroked the greased and painted skin of visiting American Indians, as did visitors to the assembly rooms in Manchester)—daring gestures at a time when evolutionary theories were challenging the old conception that the barrier separating men from brutes was immutable and that the civilized must not be contaminated by the primitive. The impulse to reach through the bars was also related to a yearning to take on a second skin, to drape an animal hide around oneself, to engage its symbolic value, to feel the stimulation of that skin, and to reinvent the self through the physicality of the exotic touch.

Given the compelling strength of the impulse to reach out and actually touch a wild skin, one wonders: what happens if the touching hand is absent? How does a person experience touch without it? These questions prompt thoughts about gazing at paintings in which the sensation of touch enters through the eyes (haptic sight). In the context of this haptic visuality, the chapter looks carefully at two Victorian paintings of human figures (one nude and one fully clothed) posed on wild animal skins: Alma-Tadema's *In the Tepidarium* (1881) and Whistler's *Symphony in White, No. 1: The White Girl* (1862). Each taps into the desire to finger the enclosed, exotic animal's hide. And each participates in the cultural anxiety emanating from the more fluid boundaries between human and beast.

Wild Animal Skins in Victorian Britain concludes with "Wild Skins and Mapping the Victorian Landscape." Because of the book's focus on exotic skins as well as on Britain's preoccupation with displays of these skins and mounted trophies, *Wild Animal Skins* fittingly ends by considering how these fixations influenced the ways in which Victorians mapped the world about them. The chapter examines the colonial adventurers' maps that take their cue from hunters' encounters with animals and the collecting of their skins; it also discusses the period's fascination with zoogeography, a cartographic genre popular in mid-century among scientists desiring to chart the distribution of animals throughout the globe. The chapter then illustrates how these views of the world influenced Hardy's *Tess of the D'Urbervilles* (in which Tess is a hunted feral being) as well as Hopkins's descriptions of landscape in his journals. Although Hopkins was occasionally indebted to the conventions associated with cartographic practices, even more than Hardy (who was indebted to the colonial adventurers' maps), he broke away from their two-dimensional, static model so that he could acknowledge the moving, feeling body and include the experience of touch. As a result, Hopkins's use of wild skins when he maps the landscape reveals a different kind of map—as unconventional as his famously idiosyncratic syntax. His maps proffer an animistic reading of nature and the world. No longer is the globe seen in terms of a captured trophy or an isolated species in a particular region; rather, the landscape is mapped through motion and texture, as well as through a vibrant intermingling of the animal and the human. (The frequent use of animal skins to describe the landscape recorded in his journals is striking.) In his maps creatures are integrated into the landscape. They no longer stand apart, isolated in their exotic otherness.

A recent experience has allowed me to understand better the analogy between wild skins and the surface of the earth. As I was finishing the book, I happened to look out of my office window and see a workman who was laying turf over the College's quadrangle that had recently been dug up to lay new pipes beneath what had once been a pleasant green space on campus. I watched the workman walk slowly toward the damaged area and noticed that over his shoulder he hauled a heavy wide swath of green turf. He had slung it over his left side as triumphantly as if he, the adventurer, had just skinned a lion he had shot and was carrying the hide to an ox-driven cart so he might transport the trophy to a ship waiting to return to England. While gazing at him, I knew then what people mean when they speak of the tundra and other surfaces of the landscape as the skin of the Earth, and I understood even better why writers, such as Hardy and Hopkins, would include animal skins in their cartographic imagination.

Preamble
Theorizing about Skin

In writing this study of wild animal skins, I have not just profited from a wealth of archival materials relating to the collecting and portraying of exotic hides in Victorian Britain. I have also benefited from recent theoretical observations concerning the significance of skin. With their attention to skin as an agent of identification and as a site of encounter, various twentieth- and twenty-first-century critics have suggested ways of viewing and analyzing many of the historical particulars I unearthed. Even though these critics, with a few exceptions, generally do not discuss animal skins but rather dwell exclusively on human skin, their observations on the import, function, and nature of the body's envelope have broadened my perspective. Although I am aware that there are significant differences between people and animals, the remarks of these critics have allowed me to think more theoretically about my subject and go beyond, though never to lose sight of, the historical data concerning the Victorians' compelling interest in the hides and feathers of birds and beasts from distant lands.[1]

Before outlining the influence of these theorists, I think it helpful initially to dwell on the connection between identity and skin in the Victorian period. For those who gazed on collections of exotic specimens, skin was the primary measure of differentiation, separating races and species, and functioned as an essential signature of being. Furthermore, recent theoretical studies have supported my perception that in the nineteenth century many regarded skin as the body's protective yet vulnerable envelope. They have also affirmed my sense that those Victorians gathering, displaying, and examining wild skins regarded skin as a surface that carries the marks of not only identity but also memory and, like a text, exists to be read. Moreover, recent thought has also allowed me to think of skin as a site of encounter and has alerted me to how the body's envelope with its multiple nerve endings orients one's way in the world. All these attributes have guided me through this study. Therefore, before going on to Chapter 1, a few pages might be in order to give some sense, though certainly not a definitive one, of the context and content of current commentaries on skin.

[1] I support Mary Midgley's contention in *Animals and Why They Matter* (1998) that although animals and humans are "incurably members one of another," the species bonds and distinctions among animals and humans are real (21, 106).

The Victorian Context

Skin has always been associated with identity. It is after all the largest and most visible organ of the body. One need, for instance, only to recall the Greek myths as well as Renaissance paintings describing or portraying the flaying of a person to realize that traditionally the removing of an individual's skin was not only an excruciatingly painful punishment but also a practice that literally stripped a person of his selfhood.[2] Throughout history, skin has served a metonymic role in representing being and has been thought of as a worn identity. Hence, in the 1676 *Notes for a Lecture on the Skin* (originally printed in Saint Bartholomew's Hospital Reports),[3] Sir Thomas Browne remarks on the fact that skin is what encloses, enfolds, or "wraps up" a person and that, as the principal and most visible part of the body, it is the means by which one recognizes "a greater division of mankind … that is into white skinned men, and negros" (108). Moreover, skin further helps identify people through its varying degrees of "coolnesse, softnesse, and smoothnesse" (110).

Although in succeeding centuries attention had been given to the health and "beauty" of people's skin (hence, a continuing focus on skin as identity), by the Victorian period this orientation had evolved into and had become absorbed by a more specialized and clinical attention. Such disorders as syphilis, herpes, leprosy, smallpox, scabies, eczema, psoriasis, cancer, and ringworm (there was a remarkable interest in this condition) were not only evidence of the individual's moral and physical condition but also a mark of the nation's wellbeing (or lack of it).[4] Integral to the period's sense of itself, the scientific study of skin was first formalized in Britain during the nineteenth century and dermatology became a specialized field of medicine. As a consequence of this more focused attention to cutaneous diseases, more and more hospitals exclusively dedicated to the diagnosis and treatment of skin disorders were established. In London, for instance, the

---

[2] For examples, see Gerard David's *Flaying of Sisamnes* (1498), and the many Renaissance and seventeenth-century representations of Apollo flaying Marsyas: for instance, Titian's *Flaying of Marsyas* (ca. 1570–1576), Giovanni Stradanus's *Apollo Flays Marsyas* (ca. 1580–1600), and Luca Giordano's early 1650s *Apollo and Marsyas*.

[3] It is interesting to note that Saint Bartholomew was reputedly martyred by being flayed alive. Outside the Duomo di Milano there is a statue of St. Bartholomew holding and draped with his own skin. The sculpture was done in 1562 by Marco d'Agrate. Michelangelo's *Last Judgment* shows St. Bartholomew displaying his flayed skin. St. Bartholomew is appropriately the patron saint of tanners.

[4] The first stanza of William Blake's late eighteenth-century poem "London" from his *Songs of Experience* comes to mind, for its lines refer to the condition of people's skin to remark upon the diseased state of England's capital and government:

> I wander thro' each charter'd street
> Near where the charter'd Thames does flow.
> And mark in every face I meet
> Marks of weakness, marks of woe.

Hospital for Diseases of the Skin (Blackfriars) was founded in 1841, the Western Dispensary for Diseases of the Skin (Fitzroy Square) was set up in 1851, and St. John's Hospital for Diseases of the Skin went into operation in 1863. This growing interest in dermatology also affected the training of medical students. Believing that skin disorders were hardly understood by either medical students or practitioners, teaching doctors insisted on more clinical instruction concerning the pathology and treatment of these diseases. In 1865, for example, Thomas Hillier complained that "Till very recently there was no special provision in any of our medical schools for clinical instruction" (1). To rectify this failing, manuals instructing medical practitioners on the subject started to appear. Among these are: Erasmus Wilson's *A Practical Treatise on Healthy Skin: with Rules for the Medical and Domestic Treatment of Cutaneous Diseases* (1844), Thomas Hunt's *A Guide to the Treatment of Diseases of the Skin: with Suggestions for Their Prevention for the Use of the Student and General Practitioner* (1865), Thomas Hillier's *Hand-Book of Skin Diseases for Students and Practitioners* (1865), and Samuel J. Bayfield's *The Skin, in Health and Disease: Being a Treatise Concerning the Nature of those Diseases Most Frequently Met with in Private Practice, with Treatment and Cases* (1867).

One striking vestige of this growing interest in dermatology can be seen today in the Gordon Museum of Guy's Hospital (London) where there are 352 wax models of skin diseases sculptured by Joseph Towne (1806–1879). From the age of 17, Towne was employed by Guy's Hospital to make a record of a multitude of cutaneous diseases so that students could examine them and learn how to recognize the various disorders. Doctors (particularly Dr. Thomas Addison [1793–1860]) who practiced at Guy's systematically sent their dermatology patients to Towne. In his basement workshop, Towne meticulously sculpted wax likenesses of their heads, limbs, or bodies that displayed both common and rare skin lesions or rashes resulting from such diseases as leprosy, Lupus Vulgaris, Popular Syphildoermus, Late Congenital Syphilis, Secondary Syphilis, Elephantitis, and ringworm.[5] Even today these are still used for teaching purposes. They are disturbing to look at, for they are not standardized models but sculptured realistic renditions depicting actual individuals: particular men, women, children, and babies. It is interesting to note that a nineteenth-century catalogue of these wax replicas illustrates how this growing interest in skin participated in the cultural compulsion to classify and put everything into proper scientific order. In 1873 the Curator of the Museum of Guy's Hospital (C. Hilton Fagge, MD) undertook to rearrange the models and

---

[5] Towne used pure beeswax or a mixture of beeswax and spermaceti wax (derived from whale oil). D. Mendes and H. Ellis's "Joseph Towne (1806–1879), Master-modeller of Wax" in *Journal of Medical Biography* 11.4 (November 2003): 212–17 gives a good outline of his work. During World War II, these wax models were originally stored in caves in Chislehurst (not far from London), but after they developed a mold (which was compromising the models' colors), they were returned to the Gordon Museum and stayed throughout the Blitz.

prepare a new catalogue that would revise the earlier one, drawn up in 1854 by Dr. Habershon, so that the newer models could be included and fuller descriptions of the contents be given. His intent was "that a student, desirous of learning from them the diseases of the skin, might more readily see what points were characteristic of each affection, and important to be noticed" (Fagge v). As if imitating the Victorian desire to order the world's flora and fauna, the Curator was significantly concerned about the accuracy of the classification and nomenclature employed for the collection. Aware of the imperfections and constant revisions which continually disrupted any attempt to organize these diseases, he despaired and decided "that instead of aiming at a strict classification of cutaneous eruptions, we may content ourselves with dividing them into what appear to be natural groups"—though, he recognized, these groups "are not mutually exclusive" (Fagge 10).

While this medical/scientific interest in these diseases reflected the general preoccupation with and frustrations accompanying classification, it also contributed to a larger cultural awareness of skin already actively at work due to the growth of empire and the consequential heightened perception of race and ethnic identity. As the largest organ readily accessible to the eye and as the most visible surface of the body, skin, in the context of empire, more than ever was a notable site of difference and functioned to register a sense of otherness. Hence, for example, the plethora of illustrations and verbal descriptions displaying the dark skin of Africans in travel accounts, adventure fiction, and novels—not to mention missionary accounts and more serious studies of faraway places—which used skin color to distinguish between the barbaric and the civilized. The color and texture of what was considered to be a being's most elemental possession allowed Westerners the opportunity to mark what was different from themselves and to distinguish one race or species from another. Without a doubt, in the Victorian period, skin was an important site of encounter with what was alien to one's self. Even though character and emotion were still partially measured through physiognomy (the shape and contour of a person's features), phrenology (tracing and interpreting the bumps on a head), and studies of expressions,[6] the bodily surface of skin was of primary importance when it came to representing and assigning or forming identity.

Skin for the Victorians was a kind of surface which conveyed a text—and often quite literally so. As such, skin continued to be an identifying signature. Tattooing became fashionable (an art prompted by more extensive travel to foreign lands where it was practiced and imitated so that in 1862 the Prince of Wales, "Bertie," was tattooed in the Holy Land). In the 1880s fashionable ladies and gentlemen

---

[6] Arthur Cheetham's 1893 *Character Reading Practically Explained* opens with the thought that "The best way of studying human nature is to study living faces and forms," such as the distance between ears, the prominence of the eyebrows, the shape of the nose and eyes as well as the strength of the handshake (1). One must not forget Charles Bell's *Essays on the Anatomy and Expression in Painting* (1806) and his *The Anatomy and Philosophy of Expressions* (1824).

secretly sported a tattoo and wealthy Londoners frequented Macdonald's Jermyn Street tattoo studio.[7] Gambier Bolton, a Fellow of the Zoological Society in London, registered his travels through tattoos executed at Macdonald's studio. (For instance, he had a falcon tattooed onto his back.) It is of interest that Bolton's 1897 article "Pictures on the Human Skin" in *The Strand Magazine: An Illustrated Monthly* makes the connection between skin and identity, for he notes the fact that sailors and travelers to dangerous and remote lands often had their names and other forms of identification tattooed in the language of their destination into their skin so that in case of death or an emergency they would be known. Skin, in this case, literally conveyed identity and functioned as a kind of passport. Significantly, at around the same time, skin functioned as a legal form of identification, for Sir Francis Galton in the late nineteenth century recognized that the ridges on the skin of fingers were permanent and lasted throughout a life, so, therefore, taking a fingerprint was soon established as a way of distinguishing one individual from another and, thus, became an important forensic tool and worked to assign identity.

During the Victorian period skin was, therefore, crucial to determining character and distinctiveness. It was a basis of classification. Although there were many medical and cosmetic texts and manuals about the care or treatment of skin and although there was much attention paid to its importance in comprehending the larger world, there seem to have been few theoretical works discussing the implications of its role as an organ of identification and orientation.[8] Neither were there more speculative studies concerning the skin's ability to reveal what lies hidden underneath its seemingly continuous envelope. The whole concept of skin as a porous boundary through which the world and the body encountered each other was understood in a practical sense, but not necessarily explored metaphysically, except, perhaps, in Victorian fiction, which often referred to a character's complexion in order to address the person's character or state of being.

Contemporary Theorizing about Skin

It was really not until the twentieth century that critics used these attributes of skin as a jumping off point to reflect upon both the physical and metaphoric nature of skin. Particularly within the last two decades, theories concerning the nature of skin have become prominent. What perhaps caught critics' imagination was the 1989 translation of Didier Anzieu's *The Skin Ego* in which this renowned French

7 For an excellent book on the history of tattooing see *Written on the Body: The Tattoo in European and American History*, ed. Jane Caplan.

8 There are always notable exceptions to such generalizations. One is Alexander Bain's work on the senses and the intellect in which this nineteenth-century psychologist is sensitive to the skin of the touching hand and its role in orienting individuals' relationship to their surroundings. See Alexander Bain, *The Senses and the Intellect* (London: John W. Parker and Sons, 1855).

psychoanalyst used the metaphor of skin to examine the interface between the inside and the outside and called upon the analogy between the protective skin and the interior psychic apparatus. His claim that the ego encloses the psychic apparatus much as the skin encloses the body was intriguing to many. Whether or not Anzieu's work stimulated a flurry of more speculative work (many critics I read acknowledge his thought), it is clear that from 2001 until 2004 a group of critics spent considerable energy considering the nature and meaning of skin. Among these, in chronological order, are Sara Ahmed and Jackie Stacey's collection of essays *Thinking through the Skin* (2001), Claudia Benthien's *Skin: On the Cultural Border between Self and the World* (2002), Claude Bouillon's *Skin: A Living Envelope* (2002), and Steven Connor's *The Book of Skin* (2004).

Although I have not engaged the psychoanalytic aspects of their work, I have benefited from their thoughts regarding skin as a place where identity resides, and as an envelope that both protects the self from and exposes it to what is outside of itself. They have reinforced the Victorian sense that skin is the metonymy of being. Furthermore, through their work I have been encouraged to think of skin as a text, a surface to be written on and painted. Moreover, these critics have allowed me more fully to realize that skin is a primary site of encounter—an important concept when one is considering the collecting, displaying, and stuffing of wild skins from exotic places.

Two of the most important books listed above are Claudia Benthien's *Skin on the Cultural Border between Self and the World* and Steven Connor's *The Book of Skin*. I am particularly indebted to Benthien's suggestive and interesting study. Benthien's notion that skin stands metonymically for the whole human being, and that it is a worn identity or a contact surface that is constantly interpreted, read, invested with or emptied of semantic meaning allowed me to approach my subject with a more thoughtful perspective and supported my sense that the Victorians regarded and used skin in much the same way. Another authoritative book is Steven Connor's *The Book of Skin*. This study contains thoroughly informed observations on the various meanings attached to skin through time. Connor's splendid sense of history as well as his own provocative reflections took me beyond the boundaries of the oft-repeated remarks about skin. His discussion allowed me not only to think more about skin and identity, but also to consider the Victorians' use of wild skins to feel some connection with and be a witness to their encounters with the exotic other. In particular, Connor's suggestion that skin has the power to restore life contributed to my thoughts about the nature of taxidermy.

More recent studies of skin have also caught my attention: Mary Flanagan and Austin Booth's edited collection of essays *re-skin* (2006), which is sensitive to skin as the ultimate site for negotiating one's relationship to the world, and *Sk-interfaces: Exploding Borders—Creating Membranes in Art, Technology and Society*, edited by Jens Hauser (2008), which through its collection of examples from contemporary art illustrates that skin can be thought of as a medium (a transition zone) rather than as a fixed barrier. Although these books are concerned with a later century, their examples encouraged me to embrace the concept of the skin as a canvas onto which one inscribes meaning and memory and contributed to my last chapter on skin and mapping the landscape. In addition, I found Anne Anlin Cheng's study of Josephine Baker (*Second Skin: Josephine Baker and the Modern Surface*, 2011) to be stimulating.[9] Cheng treats skin as a surface that serves as "a vibrant interface between the hidden and the visually available" (28). She explores Baker's fetishizing of her nakedness and her animal femininity, particularly when this early twentieth-century nightclub performer either wore her nudity as a second skin or clad herself in animal skins.

Inevitably when I was reading through these various theories of skin, the phenomenon of touch arose. This is not surprising, for skin with its multiple nerve endings is the most sensitive organ and, as such, registers the temperature, texture, density, and location of what one touches. As many critics point out, skin, because of its sensitivity, is the seat of perception and orients a person's way in the world. Taking their cue from Diderot's "Letter on the Blind" (1749) and from Alexander Bain's nineteenth-century work on touch as a source of ideas (*The Senses and the Intellect*, 1855) as well as from modern French critics such as Maurice Merleau-Ponty and Michel Serres, many critics dwell exclusively upon the haptic sense and its ability to give one access to what is outside one.[10] Skin opens our bodies to other bodies through touch. Because of this association with skin all of the theorists mentioned above spoke of the experience and qualities of touch. Because of this association, at the same time books were coming out on skin (many of which also talk about the nature of touch and the importance of the hand), there was a rash of publications concerning the experience and quality of touch. Among these are Ashley Montagu's *Touching: The Human Significance of the Skin* (1986), Laura U. Marks's *Touch* (2002), Raymond Tallis's *The Hand* (2003), Santanu Das's *Touch and Intimacy in First World War Literature* (2005), Constance

---

[9] Josephine Baker (1906–1975) was an American-born French dancer and singer popular in the 1920s. A YouTube clip shows censored footage of her infamous Banana Dance at the Folies Bergère in Paris. (For the film she is clothed rather than naked as she was in live performances.)

[10] In his "Letter on the Blind," Diderot writes:

> And I found that of all the senses the eye was the most superficial, the ear the most haughty, smell the most voluptuous, taste the most superstitious and inconstant, touch the most profound and philosophical. (As quoted in Patterson 1)
>
> I would just as soon have long arms: it seems to me that my hands would tell me more about what happens on the moon than you can find out with your eyes and your telescopes; and besides, eyes cease to see sooner than hands to touch. I would be as well off if I perfected the organ I possess, as if I obtained the organ which I am deprived of. (As quoted in Patterson 37)

In *The Senses and the Intellect* Alexander Bain writes: "Touch, considered as a source of ideas such as those, is really not a simple sense, but a compound of sense and motion; and it is to the muscular part of the sense, or to the movement of the touching organs [skin and the touching hand] that these conceptions owe their origin …" (171).

Classen's edited collection *The Book of Touch* (2005) as well as her 2012 *The Deepest Sense: A Cultural History of Touch*, and Mark Patterson's *The Senses of Touch: Haptics, Affects and Technologies* (2007). All these studies helped me extend my discussion of skin so that I might not only think of skin as the seat of perception but also consider what happens when a person extends his or her hand and touches a wild skin. In particular, I was indebted to Classen's *The Deepest Sense*, for in addition to exploring historically the various experiences of the haptic or tactile imagination as well as the importance of skin-to-skin contact, the book has a section on "Animal Skins" which emphasizes that the speechlessness of animals makes "touch an essential medium for human interaction" (93) and that "the most frequent contact humans had with animals … was not with their living bodies but with their body parts" (94). Classen's observations contributed to my understanding of the haptic sense and to my chapter on the Victorians' desire to touch animal skins. Along with other skin theorists, her sensitivity to the importance of skin as a place where identity resides and as a site of encounter have contributed in meaningful ways to *Wild Animal Skins* and its discussion of how the Victorians viewed and relied upon skins to try and comprehend the complexion of their world.

Chapter 1
Industry, Empire, Portraiture, and Skin at the Belle Vue Zoo, Manchester

> The history of portraiture is a gallery of poses, an array of types and styles which codifies the assumptions, biases, and aspirations of society.
>
> Harold Rosenberg

Introduction

In the Victorian period one of the most popular places where the public could see displays of wild skins was the zoo. Opportunities were plentiful. Thanks to the growth of empire, to improved transportation, as well as to newly opened routes to distant lands, it was now more possible than ever to collect and bring back creatures and their skins from faraway regions and put them on view before a population eager to see exotic animals and birds. Indeed, according to one commentator, in the 1860s, throughout Europe, zoological gardens were being established at the rate of one a year (Philo and Wilbert 248).

One of these establishments was the Belle Vue Zoo in Manchester. In the context of its industrial surroundings, this zoo created a set of "galleries" in which the general population could behold, with their own eyes, actual creatures from places usually familiar to them only through the printed medium. At Belle Vue, visitors could not only gaze at (and sometimes touch) the fur and feathers of exotic animals but also, in the zoo's museum, walk among displays of their stuffed skins. Belle Vue's commitment to its manufacturing and imperialistic context, as well as to its collection of wild skins, proffers a portrait of a time in which skin played a significant part in forming identity and orienting one's way in the world.

The Belle Vue Zoo

Belle Vue Zoo, founded in 1836, both celebrated and replicated the industrial surroundings of Manchester. Looking at the zoo from a distance, people in the mid-nineteenth century would have remarked upon the establishment's dazzling grounds, illuminated either by Belle Vue's own gas works or by the electricity powered by its robust generators. They also would have caught sight of the zoo's buildings, humming with the action of the steam-driven machines within. By utilizing the most up-to-date technology, Belle Vue was an extension of its neighboring manufacturing city.

For all its dramatic presentation, however, commentators on Victorian zoos have paid minimal attention to Belle Vue—indeed, this establishment tends to be overlooked by those fascinated with Victorian Manchester.[1] Compared to the more established and intellectually prominent London Zoological Society Gardens, this provincial and transparently commercial institution, once located just three miles from the center of industrial Manchester, is regarded as being relatively inconsequential. It does not rival the scientific achievement associated with the more famous London organization, for example, nor the Bristol Zoo, which emulated the Zoological Society's avid interest in taxonomy and anatomy.

Rather than compete with this kind of scientific orientation, Belle Vue chose to present itself as an institution primarily dedicated to the celebration of its industrial and imperialistic context. To this purpose, Belle Vue's grounds and exhibits were consciously constructed so as to show off the power of industry as well as to display the reach of empire. In this respect, Belle Vue projected itself as a kind of National Portrait Gallery, dedicated to promoting and glorifying Britain's achievements. Throughout Belle Vue, a series of exhibit rooms gave prominence to new technologies as well as to pictures of contemporary heroes and the spoils of empire. Wandering through the various buildings, visitors would have admired one engineering feat after another, glanced at sequences of heroic personages, and then, after reaching the menageries, as well as the Museum, they would have stood enthralled before the living representatives or the stuffed remains of creatures collected from various corners of the globe. Belle Vue was unique, for contrary to most establishments in Manchester it combined the mechanical life of the machine with the rhythms of organic life.[2]

What follows is an exploration of Belle Vue as a series of portrait galleries—indeed, as a site of portraiture. In Part One, I shall examine Belle Vue's indebtedness to its industrial and imperialistic setting in order to reveal in what ways this institution was a portrait of its age. In Part Two, I shall leave this more panoramic perspective and enter the zoo itself in order to discuss Belle Vue's analogous relationship to a portrait gallery. In the third part, I shall conclude by focusing on the ubiquitous role of skin in many of these galleries. The exhibited portraits recall the often overlooked, but important, part that skin plays in rendering identity. Given their rage to collect and mount skins, the Victorians probably realized better than we that skin is integral to identity, and, as such,

---

1 George Jennison in his unpublished history of Belle Vue, completed in 1929, introduces his typescript with the following unidentified quotation: "Since the beginning of civilization, there must have been places of amusement for the populace, yet scarcely one has left even a name behind it" (1).

2 In *Victorians and the Machine: The Literary Response to Technology* and in *Victorian Technology: Invention, Innovation, and the Rise of the Machine*, Herbert Sussman mentions that Victorian commentators oppose the organic to the mechanistic when considering the industrial landscape of Manchester. From my point of view, however, Belle Vue was an exception to this opposition. With its blend of the latest in technology with its living creatures, the contrast between a life ordered by a machine and that regulated by the rhythm of the organic was muted.

proffers an access to another being. This section introduces the main focus of *Wild Skins in Victorian Britain*, for it is dedicated not only to studying the nineteenth-century fascination with exotic animal skins but also to considering, theoretically, the meaning and significance of skin. The Belle Vue Zoo in Manchester offers an interesting context for this topic, and, as such, vividly commences the subject of this book.

Part One: A Portrait of its Age

To begin with, Belle Vue can be regarded as a vivid portrait of its time. Through its ostentatious displays of the latest machinery, gadgets, and devices, as well as through its spectacular exhibitions of mock battles, Belle Vue reflected the energy of its industrial environment and glorified the reach of empire. This perspective is visible in one of the few remaining pictures of Belle Vue: George Danson's *Manchester from Belle Vue*.

Figure 1.1 *Manchester from Belle Vue*
Source: Chetham's Library, Manchester 1861.

Danson's painting situates the institution within its manufacturing surroundings. With its recurring references to the smoke spewing from the towering factory chimneys, the image's horizon encroaches upon the foreground's sparse traces of a more pastoral era. The stunted pale-green fields are suffering.[3] The carriages and carts that wind through Danson's canvas, transport the noise and business of an industrial town into the forefront of the viewer's consciousness. Like the

[3] Charles Dickens in *Hard Times* refers to these fields as having "stunted grass" (18).

serpentine smoke from the chimneys, the crowded road coils round the canvas and seems to suffocate whatever quiet stillness had once been part of the landscape.[4]

Upon first glance Danson's picture revives a memory of the "interminable serpents of smoke" smothering Dickens's Coketown (*Hard Times* 27).[5] The painting also recalls the opening pages of Elizabeth Gaskell's *Mary Barton* in which working-class families amble through "some [idyllic] fields near Manchester" and notice, by contrast, the distant smog hanging over the manufacturing town (5). The filthy smoke and dust that literally and metaphorically smudge their families' lives seem also to hover over Danson's prospect.[6] Danson's painting, however, does not unreservedly subscribe to Gaskell's conventional comparison between the restorative countryside and the harsh industrial town, nor does it unconditionally emulate Dickens's sense of industry's intrusive harm. Rather, the painter's intentions were quite different. Instead of criticizing such a prospect, Danson, who with his two sons was a long-standing employee of Belle Vue, chose to celebrate the zoo's proximity to the industrial city. In his picture as well as in the owner's eyes, Belle Vue's grounds lead directly to the factory chimneys, to a landscape that for him and the zoo's founder, John Jennison, a former silk weaver from Macclesfield, reflects the institution's commitment to the positive power and progressiveness of industry. For them it was quite literally a "*belle vue*."

Within the zoo's boundaries, this manufacturing context was not a cause of despair but an occasion for wonder, a source of amazement. Belle Vue was a showground of nineteenth-century technology driving the manufacturing towns of northern England. Its monstrous polished and oiled machines associated with Dickens's Coketown became emblems of entertainment as well as extensions of the zoo's living exotic creatures. They were not, as they are in *Hard Times*, in direct opposition to the fanciful world of the circus. Instead, they became as magical as the horses riding around the center ring.

At Belle Vue the pulsating engines were as much on display as were the animals. This industrial apparatus transformed the zoo's buildings into kinds

---

4 The painting does not show the expanded railroad tracks, which connected Belle Vue to Manchester and beyond, and allowed cheap and speedy access between the zoo and the industrial sites. For a more particular sense of the extension of the railroad to Belle Vue, see Jennison's typescript, "The Making and Growth of the Famous Zoological Gardens, Belle Vue, Manchester," 51, and see Nicholls's *The Belle Vue Story*, 6–7.

5 For a thoughtful and informed discussion of the relation between *Hard Times* and industrial Manchester, see "The Industrial Novel and the Machine: Charles Dickens" in Sussman's *Victorians and the Machine*, 62–72.

6 Chapter 1 of Gaskell's *Mary Barton* begins: "There are some fields near Manchester, well known to the inhabitants as 'Green Heys Field,' through which runs a public footpath to a little village about two miles distant. In spite of these fields being flat and low, nay in spite of the want of wood … there is a charm about them which strikes even the inhabitant of a mountainous district, who sees and feels the effect of contrast in these common-place but thoroughly rural fields with the busy, bustling manufacturing town, he left but half an hour ago" (5).

of "Fairy palaces" or fanciful places where the working men and women (the "Hands" in *Hard Times*) could escape the monotony and harshness of their long hours in the mills.[7] Those who labored in the mills could meander through the gardens, the grottos, the refreshment rooms, and the menageries.[8] In a sense, Jennison, the zoo's proprietor, transplanted the population of the manufacturing towns to another kind of industrial site. Indeed, Jennison literally incorporated the Lancashire mills into his enterprise, for he bought and dismantled disused mills so he might use their rough timber to construct some of his own buildings. In this respect, it is significant that during periods of industrial unrest in the 1840s as well as during the American Civil War, which interfered with the cotton trade, the Belle Vue Zoo fell on hard times.[9] Industry and zoo were not alien to one another. It is also interesting to note that Belle Vue was periodically associated with the trade unionists and chartists, who marched from surrounding communities to the zoo and gathered for meals in the establishment's extensive tearoom (the refreshment room eventually seated 500 customers).[10] As Jennison's grandson remarks: "A great deal of Trade Union solidarity in Manchester has been laid by the processions to Belle Vue, and cemented over the teacups." He continues: "The Trade Union Local Branches used to say that they could not do without us, there is at least one very important socialist club, the Clarion, that has not missed its Belle Vue meeting for 20 years or more" (Jennison 80, 81).

When Jennison opened the zoo in 1836, he began with a modest collection of animals. As the years passed, he developed not only the menagerie and the grounds but also the machinery that made everything run. Jennison was intent upon transforming the zoo into a self-sufficient, efficiently managed manufacturing site. He wanted to show off the achievements of British technology. As a result, by the mid-century, Belle Vue had its own printing press, a large brewery, a bakery, a ginger beer company, an elaborate firework factory, its own butchery, a steady supply of hot water heated by steam, a dairy, its own ice production, a brick works, a Daguerreotype studio, an "Eccolobium"—a device in which eggs were hatched by steam—and, in 1897, a Kinematograph for the display of animated pictures. Wanting to be self-sufficient, the zoo also had its own staff of blacksmiths, wheelwrights, joiners, boat builders, painters, and plumbers. Furthermore, Jennison built his own gas works, which were independent of the city supply; he created his own waterworks from a deep spring so that water ran abundantly, via

---

[7] A description of Belle Vue Zoo in the nineteenth century refers to the "Fairy Dells" and the "little Fairy spots" within its grounds (Jennison 38).

[8] Belle Vue always catered to its most needy clients. A mid-century Baedeker's guide remarks that the zoo was "A popular resort much frequented by the working class" (Jennison 108). During Whitsun Week, for instance, there were reduced fees, and Sunday Schools could use part of the grounds for free.

[9] Robert Nicholls remarks on this connection between the economic well-being of the mills and Belle Vue in *The Belle Vue Story* (6, 16).

[10] In 1852, a party of Sheffield Chartists gathered in Belle Vue's tearoom.

buried pipes, throughout Belle Vue, feeding, among other things, fountains, fueled by powerful steam engines, which sprayed the water as high as 30 feet. Through the power of steam, a rackety "velocipede" (an early type of bicycle made of cast iron fixed to a round wooden platform) whizzed around at great speed; a merry-go-round, which formerly had been pushed by boys, spun propelled by steam; and, in 1894, a steam-driven "Ocean Wave" offered its spectators, who stood on a raised platform, the sensation of being at sea during a raging storm.[11]

Not only did Belle Vue embrace the latest machinery associated with its surroundings; it also pioneered certain technologies, particularly when Jennison's company reputedly introduced the first electric lights to the Manchester area. Belle Vue presented the apparatus as if it were another splendid, exotic animal for the public to view and admire. Significantly, the electric generators were initially housed above the elephant house. These two powerful entities, the mechanical and the organic, joined to represent something splendid. Eventually, in 1897, these generators were moved to another building. The 1897 *Guide* boasts of their ability to yield more than 500 horse power and fully illuminate the Gardens "with arc and incandescent light" (Nicholls 25). Crowds gathered to marvel at these generators, which seemed to be polished and cared for as if keepers were anointing the skin of some rare beast. Animal and machine were connected. The *Guide* explains:

> The great dynamo was roped off so that people could pass and admire the throbbing monster of the new age; the miracle which supplied light without fuel. Every piece of brass was burnished to a brilliant gold. The plant became a tremendous attraction, vying in interest with the strangest animals or the most alluring side-shows. (The Belle Vue Collection)

Encouraging this association between manufacturing and the animals on display, throughout other sections of Belle Vue, living creatures worked various mechanical devices and labored alongside the workmen: the elephants learned to put a penny in a box in order to receive a cake; the baboons drew water for themselves from a pump; one of the zoo's chimpanzees worked a chocolate machine; and, in 1853, elephants hauled bricks produced at the zoo's brickworks.

As a portrait of its time, Belle Vue not only incorporated the machinery of its manufacturing surroundings; it also populated its grounds with powerful reminders of Britain's economic, military, and political campaigns throughout the globe. By the early nineteenth century, the British Empire extended over more than a fifth of the world's land area and affected more than a quarter of the world's population. Belle Vue displayed the spoils of such expansion. The "Monster

[11] Jill Cronin and Frank Rhodes, in *Images of England: Belle Vue*, describe the Ocean Wave: "A circular platform, equipped like the bridge of a ship, is set among scenery painted to represent the waves of an ocean during a high wind. Round the edge of the platform is a line of small yachts. Powerful machinery makes the platform revolve, and at the same time rise and fall, giving the yachts a motion not unlike the ones they have at sea" (23).

Terrestrial Globe," constructed out of zinc and 20 feet in diameter, functioned as a grand emblem of this reality.[12] The zoo's collection of exotic beasts from Britain's colonial territories also reminded the visitor of this larger international context. Like other zoos of the time, Belle Vue was indebted to the infrastructure and routes associated with the industrial trade to gather its creatures. Shipped back to Britain, its exhibited animals and birds could be viewed as trophies of conquest. Sequestered behind bars, they stood for what Harriet Ritvo refers to as a "subjected foreignness"; they were a popular representation of human domination (*The Animal Estate* 206). In the mode of other Victorian menageries, Belle Vue served as a kind of repository, dedicated to symbolically acknowledging Britain's powerful international position ("The Order of Nature" 50).

Belle Vue's official publications encouraged this sense of dominion by exaggerating the dimensions of what was on display. Guidebooks not only highlighted the colossal size of the powerful machines responsible for running the zoo but also referred to its caged animals as "monsters." Visitors were invited to enter the "Monster Monkey House" and to gaze at the "Monster Orang-Utan," on view in Moorish-style buildings or in Indian grottos and temples; moreover, the public was encouraged to walk through Chinese-style gardens and take refreshments in a Pagoda restaurant. Reminders of a supposedly tamed or conquered empire were ever-present.[13]

Belle Vue's imperialistic portrait, though, was even more visible to its public through its spectacular evening reenactments of recent battles, which demonstrated Britain's superiority. Significant skirmishes during the Crimean War (events that were fresh in the public's memory), conflicts in India, as well as victorious scenes from the Boer War were often featured. "The Bombardment of Algiers," "The Fall of Sebastopol," "The Battle of the Nile," "The Defense of Lucknow," "The Battle of Alma," "Matabele War," were, for instance, presented respectively in 1852, 1855, 1862, 1863, 1896, and 1897. Begun in 1852, these shows distinguished Belle Vue from its competitors and insured the zoo's financial success.[14] With their display of technological wonders, they must have been extraordinary to watch. The first and least ambitious of these productions, "The Bombardment of Algiers," for instance, used "one thousand guns and fireballs, three hundred rockets, twenty

---

12 In the 1872 "Hand Book to the Zoological gardens Belle Vue," Robert Dibb describes The Monster Globe Stand as a "square wooden building with two flights of stairs, and surmounted by two tiers of Ornamental Balustrades; on the top is a Terrestrial Globe, 20 feet in circumference, which is made of Zinc, and has attracted considerable notice from its novelty and bulk" (Jennison 41). Apparently there was a similar globe in Leicester Square. See Bernard V. Lightman, "Spectacle in Leicester Square: James Wyld's Greatest Globe, 1851–1861."

13 See the nineteenth-century Belle Vue Zoo guidebooks in The Belle Vue Collection. Chetham's Library, Manchester.

14 Believing that firework displays would bring success to Belle Vue, John Jennison initiated these spectacles after he and his family had witnessed a successful firework entertainment in the Surrey Gardens, an established zoo near London.

five large shells and fifty candles" (Jennison 61–2). Featuring a different subject each season, these presentations brought thousands to its grounds.[15] Together with Belle Vue's celebration of manufacturing, these reenactments, with their glorified imperialistic agenda, added to Belle Vue's offering of a vital portrait of the times.

Continuing the impulse to demonstrate dominion by exaggerating the power and "monstrous" dimensions of its machinery and its menagerie, the official guide referred to these spectacles as "monster pictures," a term which refers to the huge panoramic backdrops initially painted by George Danson, the same artist responsible for *Manchester from Belle Vue, 1861*. Photographs of the painted sets, taken around the turn of the nineteenth century, give some sense of just how splendid these presentations must have been. A glance at the panoramas for "Matabele War," "Delhi. Storming the Kashmir Gate," and "Mons," reveals that the sets share the gigantic dimensions of the "monsters" in the menagerie as well as the generators labeled the "throbbing monster of the new age" (1897 *Guide*). The set for the 1897 production of "Matabele War" (see Figure 1.2), for example, features the Matoppo (now referred to as Matobo) Hills where the Ndebele were overcome by the British and their European-produced weapons. Men, paid to act the role of these fighters, charged each other with spears and guns; they moved through the set while fireworks, rockets, and smoke replicated the ammunitions' fire.[16] The photograph shows how realistically the painters reproduced the granite hills, the wooden valleys, the strewn boulders, the thick vegetation, and the mist arising from the swampy valley as well as from the headwater of the river. Moveable wings added to the realism, and so too did the sight of moving water and fog, produced by machinery attached to a water tower.

Danson, with his sons, rendered these canvases in a large room. When finished, they would let them down through trap doors in the floor so the backdrop could be transported to and erected on an island in the "Firework Lake." As the photographs reveal, spectators safely gazed at the spectacle from across a body of water that was enlarged over the years, the spoils from which fashioned a sloping bank on which people could assemble for a better view. A turn-of-the-century colored postcard I found at an ephemera fair in London picturing the "Open air Dancing Platform" at Belle Vue shows this viewing platform across the lake from one of Danson's painted sets.

---

15 According to George Jennison's biography of his grandfather, from 1854 onwards the spectacles, "commenced always on Whit Monday, with a display every evening that week and afterwards Monday, Wednesday and Saturday to November 5th, also Tuesday and Thursday in August, making a total of approximately 80 shows … Times of representations had to follow the daylight … [the spectacles] have been given as late as 9:45 … they were finally stereotyped at 9:15 in July and June, 9 o'clock from August Bank Holiday … and 8:30 from the end of September" (Jennison 65–6).

16 For a technical description of this and other productions at Belle Vue, see David Mayer's "The World on Fire: Pyrodramas at Belle Vue Gardens," 179–97. Women were not hired for these productions until 1936.

Figure 1.2 "Matabele War"
Source: Chetham's Library, Manchester.

Figure 1.3 "Delhi: Storming the Kashmir Gate"
Source: Chetham's Library, Manchester.

The latest in technological advances enhanced these spectacles. The monster pictures were manipulated and illuminated by elaborate lighting and firework displays. Reflected in the lake, these colored lights set the static scene in motion. Jennison, however, was not content merely to light up the panorama and thereby give an illusion of movement. As the description of "Matabele War" reveals, Jennison also wanted the thrill of adding live action to his productions. Consequently beginning in 1852 he engaged men to re-enact the battles and shoot thousands of guns, fireballs, rockets, large shells, and Roman Candles. At times Jennison even included marching columns of camels and elephants from the

Figure 1.4 "Mons 1914–1918"
Source: Chetham's Library, Manchester.

Figure 1.5 "Zoological Gardens, Belle Vue, Manchester."

zoo.[17] Weeks of training were required to prepare the animals for these events. These amazingly successful displays also featured the most up-to-date magic lantern projection, pyrotechnical and hydro engineering. Appropriately, one of the spectacles, "The Port of Manchester" (1894), glorified the latest in machinery and the triumph of modern science and technology.

As the years passed, the action and machinery grew more and more sophisticated or elaborate. With the help of moveable wings and a 50,000-foot backdrop, as well as increasingly intricate pyrotechnic devices, in 1854 the city of Moscow burned and fell before the audience's eyes.[18] By 1856, the sets had even more moveable wings or appendages composed of 10-foot square sheets supported on poles, the largest of which reached a height of 65 feet. And by 1896 still more action and additional dramatic effects vividly replicated "The Battle of Alma." Watching water fireworks, as well as firework rockets and bouquets of colored stars shoot out from a hot-air balloon, spectators below imagined themselves watching the famous battle from the deck of an Allied ship.

Part Two: A National Portrait Gallery

Far more momentous, however, than the mock battles played against a panoramic backdrop were the series of gigantic, luminescent portraits (many over 50 feet high), displaying leading figures associated with the spectacles'

---

17 Jennison's grandson describes these effects:

> The wings being from 10 to 15 feet apart give the artist great opportunities in perspective so that the distance attained in so small a space was ever the wonder of the spectators. Between the wings were trucks of various sizes from 2ft. to 12ft. wide, capable of carrying what was in effect another wing and in this way the total appearance of the panorama could be changed in a few moments; the central portion of the island could be altered at will; a large water tower behind the scenes permitted the representation of a torrent or waterfall and the series of slopes in the background could be utilized for marching columns, horses, camels and elephants. (Jennison 66)

18 The Belle Vue *Guide* for that year describes the 50,000-square-foot canvas as well as the action that transformed the scene, "The Grand Historical Picture of Moscow," into a living image:

> In front stands the sacred Kremlin; and … above all, rises the framed Belfry of Ivan Velikii. This gay and gorgeous scene exhibits a totally different aspect at dusk, when is represented that Conflagration which is said to have been more extensive, awful, and sublime than the world had perhaps seen since its foundation. Whole quarters of the city, hitherto standing, in a few minutes are seen in a blaze—the fiery contagion spreads far and wide—the upturning of iron roofs—the roar of flames—the crash of tumbling edifices—the fire-wreathed skeletons of desolated dwellings, and the boom of artillery, make up a combination appalling to contemplate. (The Belle Vue Collection)

subjects. Bringing the show to a resounding close as well as participating in the conventional connection between war and portraiture, these heroic portraits gathered the glory of industrial and imperialistic Britain into their frames. Reflecting Belle Vue's projection of itself as a site of portraiture, they collectively formed a kind of national portrait gallery, dedicated to generating the illusion or the ideal of a national community. The solid stillness of these portraits functioned as a synecdoche for the nation's ideal body.

In the early performances, a long roll of painted portraits, illuminated from behind, would terminate the show. By the 1860s, at the close of each production, parts of the central stages were removed, then wagons ("trucks") with powerful magic lantern apparatus would roll up and superimpose, onto the darkness, a colossal transparency exhibiting a picture of a distinguished person: a military leader, a reigning monarch, a politician, or an inventor. Distributed programs added to the authority of these individuals by devoting several paragraphs to the projected person's accomplishments—all calculated to prompt enthusiasm or tutor the audience's understanding of the individual, and all programmed to offer the spectator what Cynthia Freeland characterizes as a "feeling of contact with … respected and revered predecessors" (Freeland 47).

Imposing portraits of military leaders, such as Osman Pasha (the head of the Turkish troops in Crimea), Marshal Macmahon, and Admiral Codrington loomed before Belle Vue's astonished spectators. These portraits had the final word and advanced a distinctive national identity and ethos. Colossal portraits of the Duke of Wellington and Queen Victoria also functioned in these shows as a sort of semiotic system, giving credence to the phrase "God Save the Queen," which would have been written on the sky in colored fire. In yet other displays, concluding magic-lantern transparencies revealing a pantheon of the new industrial heroes, such as Daniel Adamson, Leader Williams, and Lord Egerton, attached a face and a name to the idea of the "great triumph of Modern Science and Engineering." These figures' portraits functioned as allegorical representations of desired national characteristics, such as "PERSERVERENCE, LABOUR, and INGENUITY," featured in many of the spectacles (The Belle Vue Collection).

From the turn of the century, a program describing the closing moments of "The Siege of Ladysmith" (the Boer War) gives a vivid sense of how central these portraits were to the spectacle's production as well as to the zoo's imperialistic agenda:

> Next are seen illuminated portraits, framed in
> Shamrock, Rose, and Thistle, of our victorious Commander-in-Chief.
> LORD ROBERTS, and the HEROES OF LADYSMITH and MAFEKING,
> Supported by representative figures of our home and colonial forces, and surmounted by
> the motto,
> in dazzling point and variegated fire,
> FOR QUEEN AND COUNTRY.

Then in the final glory of soaring rockets and bursting shells, a myriad
electric lights
flash forth
the honoured title: "Defenders of the Empire." (Nicholls 83)

Portraiture was not just confined to these magnificent displays. It also held a central position in other parts of Belle Vue. For this reason, in its own publicity, Belle Vue advertised itself as a place where there was a "regard to the combinations of Zoology, Ornithology, Botany, *Sculpture*, and *Painting*" (The Belle Vue Collection, [italics mine]). As a consequence, Belle Vue's nineteenth-century guidebooks emphasized the various paintings and busts of important individuals to be viewed in such places as the Music Hall where decorative shields displayed framed portraits of Sir Robert Peel and the Duke of Wellington. There the visitor could also stand before painted portraits of such notable mechanical engineers and industrialists as Richard Arkwright, Thomas Newcomen, Thomas Savery, George Stephenson, and James Watt.[19] The public was encouraged to be as attentive to these as, for instance, to a newly arrived orangutan.

Through its deliberate focus on portraiture in its buildings as well as in its spectacles, Belle Vue was participating in the mid-century vogue among weekly and monthly publications to reproduce portraits, which, in the spirit of the concluding transparencies, gave a human face to the accomplishments of the nation and represented an historical event through a recognizable human being. Thanks to the evolving technologies of print and the subsequent emergence of illustrated newspapers, journals regularly published engravings of distinguished individuals accompanied by lengthy biographies and lists of achievements—a practice that recalls the extensive explanations in the Belle Vue guides describing the figures featured in the spectacles. The *Illustrated London News* (founded in 1842) regularly mixed portraiture and biography. Other journals followed suit and often presented their own series of what they termed "National Portrait Galleries" in which accounts of leaders' accomplishments surrounded a collection of the celebrities' engraved portraits.

It is interesting to note that in 1852, when Belle Vue began concluding its productions with colossal transparencies of heroes and was simultaneously featuring portraits of significant people throughout its grounds, a movement was also underway to found a National Portrait Gallery in London so as to

[19] In the seventeenth century, Thomas Savery had initiated the steam engine; in the eighteenth century, Thomas Newcomen had created the first practical steam engine; and in the nineteenth century, James Watt, the Scottish inventor and mechanical engineer, had improved Newcomen's steam engine. Richard Arkwright had invented many of the steam-driven apparatus in the mills that had allowed for the industrial revolution. George Stephenson had, of course, been responsible for the steam engines that powered the railways so important to industry. He was known as the "Father of Railways." These were not the only portraits. In Belle Vue's music hall were pictures of Beethoven, Purcell, Hayden, and Weber.

assemble those distinguished individuals who represented the power of the nation. Partially motivated by the success of the 1851 Great Exhibition, Prince Albert had campaigned for a portrait gallery to be erected to commemorate scientists and technologists whose innovations had brought about the modern age.[20] At the same time, individuals such as Thomas Carlyle were expressing their hope that the government might finance a "Pantheon, or home of all the National Divinities" (Barlow 221). Eventually, thanks to the efforts of Lord Stanhope and to the support of Disraeli, the National Portrait Gallery as we know it today was officially established in 1856.

In a sense Belle Vue was part of this momentum. Through its series of portraits and accompanying life stories, Belle Vue assembled its own sequence of officially sanctioned celebrities. It joined the nineteenth-century popular genre of biography, which read history through the individual. Like the National Portrait Gallery, Belle Vue displayed images of the nation's leading figures so as to mark an historic moment and champion those qualities that contributed to the nation's strength. Many of these figures personified courage, perseverance, and ingenuity.

Belle Vue's portraits, however, differed from those on view in London. Rather than presenting a realistic picture of the person's actual dimensions and/or irregularities, Belle Vue emphasized the figure's heroic affect and relied more heavily upon generalized or stereotypical images of the hero. In this sense, Belle Vue was not interested in being "authentic."[21] It was not as intent as were the directors of the newly founded National Portrait Gallery upon making sure that the displayed portrait was actually rendered in the presence of the subject and hence depicted a true likeness. Rather, the directors at Belle Vue wanted their subjects to be readily identifiable and, therefore, part of an already accepted cultural language. For this reason they seem to have selected familiar portraits—those immediately recognizable to the public because they had already been reproduced in popular publications, such as the ones mentioned above. Most of these pictures reinforced a general impression of the heroic figure and functioned simply to offer a recognizable link between an historical event and a well-known name. For instance, Jacques-Louis David's portrait of Napoleon crossing the Alps, which concluded the 1872 spectacle featuring Napoleon's expedition over towering and terrifying mountain scenery, was not only already known to the public through magazine reproductions but also "heroic." (Surprisingly, in Victorian Britain, there were many admirers of Napoleon who associated his memory either with enlightened progress in opposition to reactionary monarchy or alternatively with

---

[20] For an essay on the founding of the National Portrait Gallery, see Paul Barlow, "Facing the Past and Present: The National Portrait Gallery and the Search for 'Authentic' Portraiture" in *Portraiture: Facing the Subject*, 219–38.

[21] Thomas Carlyle, one of the champions of the newly founded National Portrait Gallery, believed, as Paul Barlow points out, in "the great value of 'bodily likenesses' of historical figures, going on to claim that portraits transformed 'vague historical name[s]' into recognizable human beings" (Woodall 220).

military genius.[22]) In a stereotypical valiant pose, replete with indexical properties, the figure of a uniformed Napoleon sits astride his rearing horse. His posture and affect exude a sense of daring and control; they visually replicate the show's concluding motto: "Napoleon honour to the brave." When assembling its version of a National Portrait Gallery, Belle Vue featured portraits which conformed to cultural expectations and bridged the gap between the populace and the abstract. Like the popular and authoritative images of Queen Victoria and the Duke of Wellington, these paintings or photographs participated in what Richard Brilliant identifies as a "consensual ritual" between the subject and the political agenda surrounding the public figure.[23]

Part Three: Skin Galleries at Belle Vue

So far the focus has been on the more familiar or conventional modes of portraiture at Belle Vue: those renderings of a public figure which issue from a palette, an engraver's pen, a slab of marble, or a transparent slide. There were, however, other forms of portraiture at Belle Vue—in the menageries and the Museum—which relied, almost exclusively, upon the display of stuffed and live animal skins in order to present yet other portraits of Britain's power and global reach.

Featuring the feathers, fur, and dens of exotic captives, Belle Vue's menagerie displayed an analogous relationship to the portrait gallery. These "galleries" bring to consciousness the often forgotten but important part skin plays in portraiture. In particular, one learns that at the zoo, animal skins and feathers could and did furnish both the materials and sites of portraiture. They were not merely a spoil of empire and a reminder of how important the display of skin was in representing a picture of conquest and power. They were also yet another gesture towards Belle Vue's commitment to portraiture. Skin composed still another and more emblematic version of the portrait gallery.

The Menagerie

In the chapter "Why Look at Animals?" from *About Looking*, John Berger makes a passing reference to the parallel between the art gallery and the zoo. He remarks that "Visitors visit the zoo to look at animals. They proceed from cage to cage,

---

22 See http://www.liverpoolmuseums.org.uk for this explanation of why Napoleon was admired in Victorian Britain.

23 In *Portraiture*, Richard Brilliant quotes Harold Rosenberg (1906–1978), the American educator, philosopher, and well-known art critic, who remarks: "Portraiture involves a consensual ritual encounter which is both trusting and wary: the subject submits to the artist's interpretation while hoping to retain some control over what the interpretation will be. The history of portraiture is a gallery of poses, an array of types and styles which codifies the assumptions, biases, and aspirations of the society" (90).

not unlike visitors in an art gallery who stop in front of one painting, and then move on to the next or the one after next" (23). Indeed, people do progress sequentially from cage to cage, as if pausing in front of one picture before passing to the next—perhaps spending a longer period gazing at one particular species rather than at another.[24] Extending this analogy to Belle Vue, it is easy to imagine its animal enclosures as a series of framed pictures collectively forming a kind of portrait gallery. A description of Belle Vue's "The Menagerie" from the 1872 *Guide* helps the reader visualize this sequence of portraits. Each den is a portrait:

> Near the Monkey House is the entrance to a long building, on the ground storey of which are twenty dens occupied by South American and African lions, a Ceylon elephant, a dromedary, an ostrich, specimens of panther, leopard, hyaena, jaguar, porcupine, etc. (The Belle Vue Collection)

Berger's analogy, however, only goes so far. It overlooks an essential difference: in the zoo's galleries, skin replaces the conventional materials of portraiture. Rather than taking an incidental or even minor role as it does in painted portraits hung in places like the National Portrait Gallery, within the zoo, skin served as the primary material with which to represent and determine the subject's identity. Unlike portraits hanging in the National Portrait Gallery, there are no clothes to mark rank, no objects surrounding the subject to represent status or achievement, and no special posture to position and interpret the subject. There is just skin. An example of just how basic it is can be found in Edward Cross's *Companion to the Royal Menagerie* (1820). When describing the panther, Cross dwells exclusively upon this "curious foreign animal's" markings: "The ground colour of the Panther is a bright and beautiful tawny yellow thickly ornamented all over the upper parts of his body, shoulders and thighs with roundish black spots disposed into circles, consisting of four or five circular spots, and there is commonly in the larger circles a central spot of the same size, this species differs from the Leopard, which has very rarely any central spots" (Cross n.p.). One can imagine visitors, guided by Cross's words, carefully tracing, with their eyes, these distinguishing circles and spots on the panther's hide.

Skin as a key to identity is a familiar concept. As I have explained in the "Preamble" to this book ("Theorizing about Skin"), it is a commonly acknowledged fact that skin is the largest organ of the body, and, as such, functions as the site where identity is most readily revealed. Contemporary theorists, such as Sara Ahmed, Claudia Benthien, Austin Booth, Claude Bouillon, Steven Connor, Mary

---

[24] For Berger, though, this kind of viewing is distressing because it reminds him that, "the look between animal and man" has either been "marginalized" or "extinguished" (28). By concentrating only on a lack of visual reciprocity between the observer and the observed, Berger, however, overlooks the importance of the viewer's concentration upon the animal's skin, and fails to recognize how a consequent awareness of the skin's markings and texture allows the zoo visitor to connect with the framed wildlife on display.

Flanagan, and Jackie Stacey are among many who suggest that for most people skin offers an immediate means of recognition. As the most prominently visible part of the body, its surface effortlessly becomes a metonym standing for "the wholeness and entirety" of being (Connor 34). Whether rightly or wrongly, it instantaneously offers a clue to identity. In a sense, skin functions as a kind of map by which to find the way to the perceived being's identity and character (Flanagan and Booth 12). Like a text, its markings serve as inscriptions to be read and interpreted. This function is not difficult to understand. How often is a person immediately "profiled" because of his or her skin? And, of course, for many subjected to the challenges of living within a colonial context during the Victorian period, skin was even more primary than it perhaps is today in registering age, race, gender, and social status. Not surprisingly, for instance, one recognizes the pervasive attitude accompanying the imperialistic agenda, which reduced the African people to their skins. As Benthien points out, skin is a garment, "a worn identity" (161). It is the way of specifying and characterizing what is outside one's self.

As I have explained in "Theorizing about Skin," nineteenth-century visitors to Belle Vue would have shared this orientation and would have been all too willing to regard the skin of the creatures they were observing as essential to an understanding of them. For its public the color, texture, and patterns of the animal were fundamental, especially since the caged animals on display were left with little but their skin. Taken out of their surroundings and now devoid of place, position, social interaction with other creatures, and movement, these animals, as I have previously suggested, were merely left with what lay on the surface of their bodies. At Belle Vue, the living vibrant skins belonging to the creatures either pacing or fluttering in their enclosures served not only as an emblem of conquest but also as an index to their uniqueness (and, hence, central to their portraits).

This orientation reflected not only a general tendency on the part of people to associate skin with identity but also a cultural practice of the time in which natural historians depended upon skin in order to recognize a creature. Even though nineteenth-century scientists, such as Richard Owen, were cutting through skin and examining the organs and anatomy of various exotic creatures, they still relied upon this exterior surface to help them determine the characteristics of a particular animal or bird. A note in William Swainson's *Animals in Menageries* (1838) gives some sense of this practice. He describes a ruffed lemur by remarking on its "deep black colour"; he distinguishes it from another specimen by noting that it is not "entirely patched with large black and white spots" (40). Pages later, Swainson talks about a Bengal tiger, which "is at once distinguished from all other ferocious animals, by its formidable size, and its beautiful skin, marked throughout with narrow dark stripes upon a yellowish buff ground" (104). Prompted by this perspective as well as by their guidebooks, nineteenth-century visitors to a zoo were more than sensitive to the character of a tiger's coat, an elephant's hide, or a parrot's varied colorful markings. The stripes, spots, or density of a mammal's fur, the slipperiness or coarseness of an animal's epidermis, and the colorful arrangement of a bird's feathers snagged the sightseer's eye, stood metonymically for the creature on display and literally portrayed difference. Skin composed the creature's portrait. The Belle Vue narrative (as well as other zoo guides) encouraged this orientation by remarking on an animal's skin and by commenting upon such features as the thickness of a rhinoceros's hide, the exquisite markings of a Bengal tiger, or by noting the visible secretions oozing from a young hippopotamus's pores. An 1896 Belle Vue Zoo guidebook, for instance, notes that the hippopotamus is "much hunted for the tough hide" and that the rhinoceros carries a "thick and almost impenetrable hide" (The Belle Vue Collection). Similarly, other descriptions focused the visitor's attention on skin by commenting, for example, on the necessity of anointing an elephant to preserve the lubricity of its hide.

When looking at these various animals, however, the Victorian public might have also been thinking about the distinctive appearance of their own skins, which distinguished them from the beasts enclosed in the menagerie. They would have been considering their own portraits as well. The sight of an animal's patterned fur, a rhinoceros's overlapping hide, or a lizard's slick scales, in contrast to the viewer's own relatively smooth epidermis, not only created a gallery of exotic portraits for the zoo visitor but also presented observers with one of the most facile means of differentiating themselves from the beast. Anxious not to be associated with any sort of primitive otherness, to keep it alien, they focused on the creature's skin and registered its boundaries so as to secure a more acceptable or "civilized" portrait of themselves. In a culture reeling from Darwin's *Origin of the Species* as well as from a general understanding that the characterization of species was under a continuous process of revision, visitors to Victorian zoos would have been more than involved in these kinds of negotiations, and would have been anxious about the permeable boundaries separating man from beast.[25] While gazing at an animal, they would have often been consciously wondering about their relationship to the animal world, and would have been sensitive to the vulnerability, as well as ambiguity, of their own skin—does it really distinguish (and therefore create a boundary) between the person and the beast? They would have apprehensively deflected their gaze back to themselves. Their "civilized" portraits of themselves were not as secure as they might wish.

Most visitors to the zoo would have liked to regard their own skin as a kind of protective envelope that visually separated them from and obfuscated any confusion between themselves and the animal world.[26] Lacking any superficial resemblance

[25] This concern about the permeable boundary between man and beast continues. In *re:skin*, the editors, Mary Flanagan and Austin Booth, recall a recent work in which Sara Diamond photographs herself wearing furs, the patterns of which inspired Diamond "to question identity, species, behavioral practices, and the boundary lines between 'animal' and 'human'" (8).

[26] Skin not only provides an identifying surface, it also serves as a protective envelope; it shields one against what is external to it (Bouillon 31). Like an envelope, skin ostensibly contains the individual and, in such a way, separates one being from another.

to one another, a comparison of the two sets of skin reassured spectators that they were distinctly different from the beast. How could I, with my smooth, rosy skin be mistaken for or related to the hairy ape? In this role skin became more than just a surface or living envelope. It also acted to separate the body or the self from what lies outside of it. As Bouillon, Ahmed, and Stacey observe, skin "forms a protective envelope and constitutes the first line of defense" (Bouillon 31) against a threatening external world; it covers the body and "protects us from the other" (Ahmed and Stacey 1).

However, as theorists, such as Connor, are quick to mention, skin is not necessarily a sealed, airtight border. Broken up by orifices, it is not continuous. It "is vulnerable" (65). The protective envelope can be penetrated or broken. Furthermore, being porous, skin can absorb what is supposedly kept outside its margins. As a result, even though skin works as a boundary, it also functions as the place of encounter and ends up being "the ultimate site for negotiating our relationship with the world" (Flanagan and Booth 3). This reality was threatening to many who feared that the boundary between man and beast was not as fixed as they might wish.

Inevitably *Punch* satirized the consequent insecurity radiating from this vulnerability and published a series of cartoons, set in the zoo, which mocks the public's general desire to hold itself above the lower orders. The portraits in these cartoons reveal the shifting margins between man and beast; they unfasten the so-called secure envelope enclosing the civilized body. In a majority of these, the humor appropriately depends upon a comparison between the skin of the creature and the person regarding it. In 1865, for instance, *Punch* printed a sketch of a dignified gentleman gazing at a penguin in its enclosure. This visitor, appropriately named "Aptenodytes Pennantis, Esq.," much to his discomfort and astonishment, discovers that his outward appearance and skin coloring almost exactly resemble that of what he is observing. He exclaims, "Dear Me!" The penguin is not so pleased with the revelation either (48: 181).

Linley Sanbourne's series of cartoons in *Punch*, "Designs after Nature," also mocked this uneasiness by dressing or enveloping fashionable ladies in the skin of animals—an exercise that not only devastated the latest in dress but also poked malicious fun at those who regarded themselves as superior to the monkey tribe as well as to the other so-called lower orders. As Susan D. Bernstein observes, Sanborne was reflecting "a cultural preoccupation with taxonomies of nature," and was speculating about the margins between humans and other animals (*Victorian Animal Dreams* 65). The result is that in Sanborne's cartoons a lady draws dangerously close to revealing her animal nature when she dons an animal's skin. For instance, an October 12, 1867 sketch, "Coming to the Point," shows a young woman, adorned, from head to foot, in porcupine quills, mincing down a fashionable London street. Prickly, she is "The next sweet thing à la porcupine" (53: 152).

APTENODYTES PENNANTIS, ESQ.

(A Sketch taken in the Zoological Gardens).

Figure 1.6 “Aptenodytes Pennantis, Esq.”
Source: Punch 1865.

"COMING TO THE POINT."

THE NEXT SWEET THING À LA PORCUPINE.

Figure 1.7 "Coming to the Point"
Source: Punch 1867.

A turn-of-the-century photograph in the Belle Vue Collection unintentionally, I think, invites a similar perspective. The image, printed below, portrays a group of people looking through the bars of the pelican enclosure. Fascinating are the one or two ladies, wearing feathered hats and frilled blouses, who are pressing up to the cage. They appear more bird-like than the rather sedate pelicans.

Figure 1.8 "Pelican Enclosure"
Source: Chetham's Library, Manchester.

Other cartoons in *Punch* also reflect this anxiety by ridiculing visitors who are blind to their own resemblance to the animals behind the bars. In "At the Zoo" (August 3, 1867) two gentleman, one resembling a hippopotamus and the other a giraffe, stand outside an enclosure and wonder, in a self-satisfied manner, just how difficult it must be for a hippopotamus to walk and the giraffes "to maintain their Perpendicular" (53: 50).[27]

[27] In yet another example from a March 6, 1869 issue of *Punch*, the worry about the difference between animal and beast takes a humorous twist. "A Gentle Vegetarian" depicts a hippopotamus with a broad open mouth gazing at a petite, demure young woman standing before its cage. The hippopotamus proclaims: "Morning, Miss! Who'd ever think, looking at us two, that you devoured bullocks and sheep, and *I* never took anything but rice!" (56: 90).

Figure 1.9 "At the Zoo"
Source: Punch 1867.

It is interesting to note that at a time when the public feared that memories of the primitive beast might still be lurking beneath the so-called protective envelope of civilized skin, Victorian portrait painters were sometimes instructed to be careful not to give, inadvertently, a hint of man's simian origins. John Burnet, for example, in his 1850 *Practical Hints on Portrait Painting*, recalls an extraordinary mandate from the pen of Sir Charles Bell. Amidst a conventional set of rules concerning the painting of a portrait, Burnet instructs his readers to position the hands of their female subjects so that "the fingers … bend slightly outwards" to reveal the proper flesh tints. This position, he believed, was necessary in order to set these graceful beings apart from the "brute creation" (the chimpanzee) that bent its fingers inwards so as "to fit them for clinging to trees" (Burnet 35).[28]

[28] Not surprisingly religious texts also reflected this underlying anxiety concerning the ambiguous boundaries separating the animal and the human. For instance, in *The Zoological Garden: An Invitation to the Youth of Both Sexes to Contemplate Animated Nature* (1859), Gustavus Canton points toward the various zoo animals, both stuffed and alive, and sternly lectures his young listeners: "One thing more I must tell you of, that disgusts me in a still higher degree than the Ape does, and this is, when a human being descends so low as to imitate the animal in his brutal sensuality, and to make himself, as it were, the Ape of the Ape" (92–3). Other examples registering a horror of the simian's lurking

Burnet's advice anticipates a particular episode in Robert Louis Stevenson's *Strange Case of Dr Jekyll and Mr Hyde* in which the primitive ape-like Hyde breaks through the borders of the doctor's civilized exterior skin.[29] In the scene, Dr. Jekyll goes to bed thinking he has reverted to his own respectable self, but when he awakens, to his horror, he glances at his hand and discovers, instead, the shaggy and gnarled skin of an ape's limb emerging from under the bed covers. The mirror in his room confirms that Dr. Jekyll has lost control over his primitive being and has woken as Mr. Hyde. No longer is the ape-like, hairy "hide" enclosed within the boundary of the respectable doctor's epidermis and, therefore, out of view. The secure envelope is torn.[30] The anxiety informing Burnet's instructions to the portrait painter is realized. The skin of the brute's hand has broken through the barrier of the civilized exterior envelope to expose its primitive character. The delicate balance between man and beast is shattered.

The Belle Vue Museum

Once members of the public had completed viewing the live creatures framed within their dens, they would have walked either upstairs or through the zoo's grounds and eventually reached the Belle Vue Museum, in which stuffed skins and collections of specimens composed a still-life menagerie, reminiscent, perhaps, of a sculpture gallery. This attraction offered an additional emblematic site of portraiture.[31] At Belle Vue, as well as at other museums of the period, the skin of a deceased creature provided the envelope, which a taxidermist filled as if to revivify the animal or bird for visitors to gaze at.[32] Because skin had once contained life, it was thought to have the power, virtually, to bring the living creature back.[33]

---

shadow are not hard to find. For instance, nineteenth-century followers of physiognomy were habitually echoing expressions of what Susan D. Bernstein calls "ape anxiety" and noting that the preferred type of nose was that which "displayed a significant distance from any resemblance to an animal" (Cowling 79).

29 For a helpful article on the whole question of "ape anxiety" in the Victoria era, see Susan D. Bernstein, "Ape Anxiety: Sensation Fiction, Evolution, and the Genre Question."

30 With reference to this argument, it is interesting to note that the structure of Stevenson's *Strange Case of Dr Jekyll and Mr Hyde* is based upon one envelope after another being opened and exposed to reveal a narrative.

31 In the Victorian period, a museum was usually an essential component of a zoo. Belle Vue was no exception.

32 To understand the art and practice of taxidermy in the Victorian period, one of the most helpful books is *A History of Taxidermy: Art, Science and Bad Taste* by P.A. Morris.

33 Since mortality in the zoos was a significant reality, these skins were readily available. A few lines from an 1830 poem bluntly refers to this reality: "We show them first the nat'ral way, / And then our beasts we *stuff*" (*The Zoological Keepsake* ix). Poor diets, cold temperatures, festering skin abscesses, growths on the feet, fractures from falls, infections, and wounds from fights were among the many causes of deaths in menageries. Jennison's ledger book records Belle Vue's particular losses and additions. It lists the

Though Belle Vue's Museum reputedly attracted attention from "the first naturalists of the day" (Jennison 38), it was a poor imitation of the Zoological Society's more professional and extensive efforts.[34] In spite of its more modest character, however, visitors willingly spent an extra penny so as leisurely to examine the exhibited stuffed skins and skeletons of wild animals.[35] A significant number of this public would have possessed a profound regard for natural history and would have even learned the art of taxidermy. Like the elderly self-educated spinner Job Legh in *Mary Barton*, many would have been "warm and devoted followers" of the "more popularly interesting branches of natural history." They would have known "the name and habitat of every plant within a day's walking from their dwellings" (Gaskell 39). Gaskell's Job Legh, for instance, meticulously inspects "every new specimen with real scientific delight" (39). When he acquires a "flying fish" from a close family friend, Will Wilson, he is able to recite its proper Latin name: "It's the Exocetus; one of the Malacopterygii Abdominales" (153). The name is not nonsense; it is technical. Legh, it seems, knows Georges Cuvier's classification system, most specifically Cuvier's research on fish published at the beginning of the nineteenth century, which classified fish according to the nature of the creature's rays on the dorsal and anal fins. As Léon Foucher observed in 1844: "Science which is so often developed by the progress of industry [had] fixed itself in Lancashire" (Kargon 3). It is common knowledge that in Manchester, a number of mill owners and enlightened businessmen endeavored to include their employees in recent discoveries in technology and science. For instance, in 1821, a group of entrepreneurs established the Manchester Mechanics' Institute. In addition, more established organizations, such as Manchester's exclusive Natural History Society, occasionally invited non-subscribers and working men to special exhibits.[36]

---

accidental deaths from animals injuring themselves or eating each other (such as the time an African snake, donated from the Soudan Campaign, consumed a lizard). It also enumerates the losses from disease, rat bites, improper foods, cold temperature, as well as from malicious visitors—"Some lad killed a Coypus [a river rat] with iron bar … cages must be made fool proof as well as secure" (The Belle Vue Collection). The book not only names diseases (such as bulbous foot in penguins) and describes cures but also notes how best to preserve the skin of recently expired specimens. One entry suggests rubbing in "plenty of powdered saltpeter" and gives the names of skin dressers and taxidermists.

[34] For further information about the Museum at the Zoological Gardens in London, see P. Chalmers Mitchell's *Centenary History of the Zoological Society of London*.

[35] A charge of a penny was added when the Museum moved from the center of the island to new and larger quarters at Belle Vue's main entrance in 1856.

[36] It should be noted, however, that in *Science in Victorian Manchester*, Robert H. Kargon also points out that there were societies which fought against democratization. The upper middle class in Manchester was sometimes unwilling "to support financially and by personal attendance any organization or institution which would provide for the mixing of classes. The laboring classes were effectively excluded from the benefits of the Royal Manchester Institution" (19–20).

Significantly, a sufficient number among Belle Vue's public, including people such as Job Legh, not only would have possessed scientific knowledge about the exhibited specimens but also would have been cognizant of the skills needed to perform the art of taxidermy. Like Legh, they would have filled their shelves with curiosities from foreign lands in order to fashion their own miniature museums, and they would have looked appreciatively at the Museum's creatures, for these were more professional renditions of their amateur and less ambitious attempts at home.

Individuals who wandered through its rooms would have also been fascinated with the Museum's random group of curiosities, which included, among other things, a boa constrictor, a small geological collection, random spoils of imperialistic campaigns, and, significantly, various "Busts of Heroes" (see poem quoted below). This display of sculptured heroic figures mingled with the resurrected skin of the deceased creatures (the taxidermy models) to transform the Museum into yet another portrait gallery at Belle Vue. Here a person could not only regard the sculptured face of a triumphant national figure but also appreciate the taxidermist's skill and gaze at a skin that had been cured, filled, and shaped to portray the once living creature. As it had in the menagerie, empire once more fused with natural history. A verse from an 1861 poem catches one person's enthusiasm for "this splendid museum"—a site, according to the founder's grandson, which "made a far larger appeal than one could imagine probable" (Jennison 106–7). In fact, so popular was the Museum that in 1856 it moved to larger quarters at the zoo's main entrance:

> THE SPLENDID MUSEUM! Presents such a mass,
> That a few of the principal features we class;
> But a volume would scarcely suffice to convey,
> The wonderful objects that meet you each way.
> There are Birds from all nations—all colours—and size,
> And Wild Beasts in numbers with dark flashing eyes:
> *The King of Oude's Tiger*—for fighting so famous,
> With the skull of a wonderful Hippopotamus;
> There's the Lion that once with the Dog shared its Den,
> And Trophies likewise used by wonderful men;
> There are Fishes—and Insects—and Monkeys—and Cats,
> A Duck with four legs, and most strange looking Rats;
> There's old fashioned Armour, and Indian Canoes,
> And Busts of Heroes—and spoils from our foes;
> Boa Constrictors—and Adders—and Lizards—and Snakes,
> And War Implements of most curious makes;
> In fact such a treat for a penny's in store,
> You should pause not a moment but into the door.
> (The Belle Vue Collection)

A more sober description from an 1891 guidebook specifically refers to the Museum as a "gallery":

> Here in a large gallery are [*sic*] a miscellaneous collection of Birds, Animals, Reptiles, etc., almost all which have lived in the Gardens. At the end of the room is a Skeleton of the celebrated performing Elephant "Maharajah," late of Wombwells, which at the dispersal of that renowned collection in the year 1872, was purchased by Messrs. Jennison, and lived ten years in the Gardens. (The Belle Vue Collection)

To reach Belle Vue, this elephant had walked to Manchester from Wombwell's Menagerie in Edinburgh at "abt 20 miles a day" (Jennison 82).

A photograph in The Belle Vue Collection, Chetham Library, taken at the end of the nineteenth century gives some sense of what it must have been like to step inside the Belle Vue Museum, a building 75 feet long, 25 feet wide, and 21 feet high. This image reveals just how important skin was in creating the portraits of the displayed creatures within this gallery. Behind the Keeper (probably Mr. Caruthers) stands a mounted tiger—its striped skin, prepared by a taxidermist, lurks behind the railings. Perhaps this is "King Oude's Fighting Tiger," mentioned in the above poem, famous for its staged battles with other wild beasts, particularly other tigers, in the court of Nusir-V-Deer, the King of a small territory close to Calcutta.[37] Or perhaps it is the tiger well known, in 1863, for its double stripes. In addition, in front of the tiger, between the elephant's front legs, recline the stuffed remains of a dog and a lion. When alive, the two had befriended each other and shared a cage at Belle Vue.

At first glance the skins of these specimens appear to be secondary to the spray of bones, which overwhelm the photographic image. The hippopotamus's gaping jaws, perched menacingly on a pedestal, and the towering skeleton of "Maharajah" possess the frame. However, a second, less fleeting, glimpse diminishes the predominance of these skeletons and pulls the attention back to the more essential presence of skin. To the left of the Keeper's arm is a piece taken from the elephant's trunk. Like an envelope waiting to be filled, this sample of skin hangs near the beast's tusks. Although barely visible in the photograph, this telling detail essentially affirms the elephant's reality, and, more than anything else on display resurrects "Maharajah" before the viewer's eyes. The Keeper of the Belle Vue Museum has suspended this sample of skin from the elephant's trunk not just to recall the sight of the animal and its once seamless surface but also to remind the visitor of the skin's texture, its wrinkled, hairy exterior. Texture holds memory;

[37] An account of these fights can be found in *A Private Life of an Eastern King by a Member of the Household of his Late Majesty, Nussir-V-Deen, King of Oude* (New York: Redfield, 1855), 180–82. King Oude's tiger had reputedly killed 30 lions and one more after the animal was transferred to the London Zoological Gardens. See also *Harper's Weekly*, June 11, 1859 issue.

Figure 1.10 "Elephant Skeleton in Belle Vue Museum"
Source: Chetham's Library, Manchester.

traces of what one has touched remain in a person's consciousness. As Connor remarks in his study of skin, "the world presented to my touch imprints itself on me" (34). People looking at this sample would have recalled the fact that they once had actually been able to finger and stroke that very trunk, to be in direct contact with it, especially those who had fed the elephant buns, fruit, and cake (a common practice) or ridden upon its back (elephant rides were a popular feature of Belle Vue).

More than the skeleton's bones, memories of feeling the skin's texture revive the life and movement that had once resided within its envelope. Through the memory of touch, this piece of skin functions as the passage to the elephant's existence. As if acknowledging this reality, a leading nineteenth-century taxidermist, Captain Thomas Brown, suggested to conservators of museums that they should distinguish between a two-dimensional portrait and a preserved skin. Brown believed that because of its three-dimensional and uneven surface, an animal's skin brings viewers closer to what they regard. He maintained: "Although good drawings and engravings will give us a perfect knowledge of the general appearance of animals, still they are deficient in many particulars; for by them we cannot be made acquainted with the texture of skin, nor the structure of the hair or feathers" (Brown 1). For most observers, except perhaps the anatomist, a skeleton is devoid of such cues. The skeleton alone is not enough; its bleached, lifeless bones are void of any fleshly reminder of the living beast.

Skin's essential role in opening up a right of way to portraiture is also rendered obvious in other nineteenth-century natural history sites. For instance, in 1899, at a Zoological Society meeting in London, the only existing slice of skin from a giant ground sloth, found in Patagonia, created immense excitement. Next to the extinct animal's existing skeletal remains, this portion of the sloth's hide infused life into the bones and offered the Society members a more convincing access to what once had been a living, textured creature (Scherren 219). For a similar reason, earlier in the century the remnants of a mammoth's skin, discovered in the melting Siberian ice, were shipped off to the Museum of the Royal College of Surgeons. Albeit spoiled and "offensive," this dry and hard fragment of skin was thought to be considerably closer to the "real thing" than the skeleton.[38] Though imperfect, this relic represented the creature it had once enveloped. In another example from the period, the Curator of the Zoological Society Museum depended upon a fragment of damaged skin to introduce his colleagues to a newly discovered species, a Red-eared Monkey. The dependency upon these pieces recalls Benthien's remark about the role of human skin following the death of a person: "The skin (and not the soul) is what remains of the person after death … and makes a person identifiable" (96). Skin is crucial to being.

Conclusion

Though not as dedicated to scientific research as the prominent London Zoological Society, Belle Vue was devoted to projecting a vibrant portrait of its time. As an expression of this mission, Belle Vue created an assemblage of portrait galleries, celebrating the power of industry and the reach of empire. These galleries displayed a series of pictures, among which were those which depended not only upon the more conventional forms of portraiture to be found in institutions such as the National Portrait Gallery but also upon the medium and the display of skin. At Belle Vue, skin was more than just a tangible souvenir of conquest—a trophy of imperialism and occupation; it also functioned as a compelling ingredient in portraiture. In particular, as I have discussed, the portraits displayed in Belle Vue's menagerie and Museum were vivid reminders of the ubiquity and importance of skin in the rendering of identity. Their contents remind one that skin can be a significant site of representation and interpretation. As the previous pages suggest, skin, the largest organ of the body, can function as a surface map pointing the way to identity; its patterns and texture mark type, individuality, and difference. Skin can also serve as a protective living envelope, which forms a boundary between one being and another, defines what it encloses, and defends that being from what

---

[38] This piece of skin when first brought to the museum was described in *The Pictorial Museum of Animated Nature* as being "offensive" (86).

lies outside its borders. Paradoxically, though, this boundary or protective envelope is not invulnerable. Skin is often exposed to the influence of what lies outside its supposed limits. It does not necessarily keep away an alien other as one might wish. As demonstrated in the section on the permeable boundaries separating man and beast, skin does not always function as the reassuring envelope of the body.

I conclude with a humorous incident, recorded in Jennison's history of Belle Vue. The episode offers a whimsical reminder of the powerful association between skin and being. Young James Jennison, the son of the founder of Belle Vue, apparently caused "absolute terror" when, with a bear skin draped over himself, he crawled among the zoo workers sitting around a fire in the kitchen. He remembers: "my word, the men were frightened, and didn't I get whacked for it by my father" (Jennison 128). Though hardly a Heracles, who donned the skin of the vanquished Nemean lion, the young Jennison proved that it is possible to adopt the strength of another by stepping into its skin. Skin was all the young boy needed.

The power of skin to portray identity in the above anecdote recalls yet another moment at Belle Vue. In addition to recruiting live actors to play the parts of soldiers or frightened natives in the Belle Vue spectacle or "Monster Pictures," sometimes the management, almost parodying the task of a taxidermist, stuffed sacks (these sacks represented the envelope of skin) and made effigies of the living so that they could pierce a soldier's "flesh," blow a person to pieces, throw human remains into the fire, and toss the enemy over the wall. A description of the 1855 production replicating the storming of the Malakoff and Redan (a famous battle in the Crimean War) mentions these effigies: "The Russian, French, and British soldiers were shown in fatal conflict, with an appearance of reality that often elicited from female spectators audible exclamations of alarm, particularly when an occasional *stuffed* soldier was hurled over the battlements" (The Belle Vue Collection). No matter what form it takes, skin is convincing—even from a distance.

The power of skin in representing identity and orienting the self is something not to be overlooked. With this understanding in mind, the remainder of this study will explore the role of animal skins in Victorian culture; it will focus on the reality of collecting animal skins from the reach of empire, the centrality of skin in Victorian animal portraiture, and the desire to touch wild skins. The book will conclude by exploring the literal and metaphoric function of skin in mapping the nineteenth-century landscape.

An Afterword

Before passing on to the next chapter, readers might be interested in knowing what eventually happened to the Belle Vue Zoo in the following century.

With the passage of years, the Belle Vue Zoo was necessarily subjected to the vagaries of time. In the twentieth century, the Museum, discussed above, suffered from neglect. A popular interest in acquiring natural history specimens had declined—most definitely, taxidermy had gone out of style.[39] Eventually in 1937 the Museum closed, and the few unspoiled specimens (including taxidermy items) were sent to the Manchester Museum.[40] The menagerie, on the other hand, continued, but, as might be expected, in the 1940s the ravages of war took their toll and compromised the menagerie's former aura of imperialistic control and authority.[41] No bombs fell directly on Belle Vue (there was a plan to arm guards in case dangerous animals got out during a raid). After the War, the zoo's menagerie recovered in number and quality; the public flocked back.

Intermittently there seems to have been an impulse to retain an attachment to the previous century's glorification of empire. For instance (with the exception of the war years), Belle Vue persisted in producing spectacular firework displays, many of which repeated those created during the 1800s. As if acknowledging the decline of empire, however, these shows did not realize a profit. Consequently, from 1956 until the late 1960s, Belle Vue merely presented what the management appropriately labeled "Ordinary displays" (Nicholls 67).

To some degree, Belle Vue continued to celebrate its alliance with the mills and manufacturing. In 1932, the zoo hosted a spectacular Cotton Pageant, which, with the support of over 12,000 people, celebrated King Cotton and presented scenes replicating "The Age of Invention," "The Progress of Transport," "Lancashire at Work," and "Lancashire Cotton for the World" (Nicholls 33–4). Later Belle Vue's extensive exhibition hall housed the annual Textile Recorded Exhibition (the largest in the world), which presented the latest in textile machinery. Belle Vue also continued to be a place delighting in the latest technological developments and engineering devices: such features as its speedway, its motorcycle races, its stock car racing, the entertainment park's thrilling rides, its 60-foot-high cable car, as well as its "V2 Rocket Man," who shot out of a barrel at 120 miles per hour, carried on the earlier excitement associated with such highlights as the former steam-driven velocipedes or "The Ocean Wave." These and other such devices recalled the exhilaration once experienced in the presence of the various steam-powered machines and "monster" generators that had lit, heated, and empowered Belle Vue in the nineteenth century.

---

39 Today museums often apologize for putting such specimens on display; many people now feel squeamish viewing a stuffed creature, so much so that contemporary artists occasionally use taxidermy for its shock value.

40 The skeleton of the elephant "Maharajah," mentioned in "The Splendid Museum," quoted in the text, can still be viewed at the Manchester Museum.

41 Many of Belle Vue's buildings were used as war offices. Animal exhibits were severely compromised. Nicholls's history describes how the keepers (many of whom were substitutes for the regular ones who had been sent off to war), in spite of special rationing for the zoo, struggled to feed the birds and animals. The birds in the aviary died from lack of seed; the sea lions suffered from eating strips of horsemeat soaked in cold liver oil, and the monkeys existed on potatoes.

The older Belle Vue of the Jennisons, however, ultimately disappeared. Caught in the demolition of the passing years, the old pumphouse, for instance, which the Jennisons had proudly erected, was knocked down in 1961, and so were many other features of that time. The management also changed character. By degrees a corporate leadership replaced what had been a local family business. The animals were gradually sold off in 1977 to make more room for a profit-making leisure complex. But after the post-war boom slackened, the plan for such a facility failed, and in 1982 Belle Vue closed down completely, with the exception of its exhibit halls, which remained open until 1987. It is perhaps appropriate that among the few material items now left are the skeleton of "Maharajah" and, significantly, a few remaining stuffed animals and study skins in the Manchester Museum.[42] Throughout much of its history, Belle Vue had been in the practice of sending animals which had died over to the Manchester Museum where zoologists and taxidermists would either mount the skins or treat the hide so that it could be kept behind the scenes as a "study skin" for those interested in taxonomy. In turn, Belle Vue benefited from the expertise of those associated with the Manchester Museum and learned, for instance, the reasons why a particular creature had died and how better to care for such an animal in the future. Although Belle Vue lives on in the archives at Chetham's Library, Manchester, the skins at the Manchester Museum also significantly contribute to its memory. Skin, it seems, plays an indispensable part in rendering a portrait of an age. It composes the afterlife of not only a living creature but also the zoo itself.[43] In a sense, skin has the last word.

[42] Belle Vue, I hasten to add, also continues to "exist" in its archives housed in the Chetham's Library, Manchester. There the zoo's ledger books, letters, notes, photographs, maps, accounts, and typescripts paint a remarkable portrait of a place once dedicated to the celebration of its industrial milieu and to a display of empire.

[43] For an interesting essay on one of these skins, see Sophie Everest's "Under the Skin: The Biography of a Manchester Mandrill" in Alberti's *The Afterlives of Animals*, 75–91. She writes about a male mandrill, which lived for a while at the Belle Vue Zoo. In 1909 the mandrill died. Its body was donated to the Manchester Museum where it was skinned and saved as a "study skin." This skin joins hundreds of others kept in storage areas of the Manchester Museum.

Chapter 2
A Skin Disorder

> Natural History is far too much a science of dead things: a *necrology*. It is mainly conversant with dry skins furred or feathered, blackened, shriveled, and half-stuffed; with objects … with uncouth forms, disgusting to sight and smell … These distorted things are described; their sides, plates, feathers counted; their forms copied, all shriveled and stiffened as they are; their colours, changed and modified by death or partial decay … the whole is enveloped in a mystic cloud of Graeco-Latino-English phraseology (often barbaric enough);—and this is Natural History.
>
> Philip Gosse, *A Naturalist's Sojourn in Jamaica*, 1851[1]

Introduction

In 1836, as Animal Preserver for the British Museum and the London Zoological Society, John Gould skinned and stuffed 400 birds; when the 13th Earl of Derby died in 1851, he left 2,500 exotic bird skins to the city of Liverpool, and by the time Alfred Wallace departed from the Malay Archipelago in 1862, he had accumulated over 8,000 skins. On a smaller scale, during the latter half of the nineteenth century, the Rev. Walter T. Currie, like many other missionaries throughout the world, shipped back numerous samples of skins and hides from West Central Africa so that they could be used for scientific purposes as well as for display in museums (Barringer and Flynn 156). One wonders: why skins and why so many?

Such startling numbers occasion a careful look at the persistent interest in collecting and classifying specimens gathered from the reach of empire. Most commentators equate this eager pursuit with Britain's desire to control foreign territories. Sometimes this perspective is appropriate, for, as I point out in the previous chapter, a caged lion, a skinned tiger, or a stuffed gorilla displayed in nineteenth-century menageries and museums did serve as tangible proof of Britain's mastery abroad. These "tamed" animals functioned as essential features of the larger imperialistic portrait. I worry, though, that such an equation is, at times, misleading. However accurate in certain contexts, a habitual and exclusive association of natural history collecting with the imperialistic yearning for power and possession is ultimately insensitive to the actual experiences of amassing foreign specimens. The paradigm is reductive and hence limiting. As a consequence, by focusing solely on the idea of dominance, those who agree with critics such as Harriet Ritvo or Sarah Gregory Kohlstedt and their insistence on the imperialistic control over the natural world fail to acknowledge the messiness

1 As quoted in Barber 41.

of empire.[2] I suggest that collecting was a much untidier process than most critics acknowledge.

Even during the height of Britain's power, no matter what steps were taken to subject alien species to the authority of the metropole, either by means of taxonomical arrangements or by enclosing exotic specimens within "tame" or civilized spaces, accident, mortality, contagion, and confusion inevitably threatened illusions of possession and disrupted fantasies of conquest. Mastery was never fully realized. For Victorian naturalists, Walter Benjamin's sense that every passion for collecting borders on the chaotic was unmistakably a reality.[3] They too opened their packed crates (for them, filled with natural history specimens) and were aware of the perplexity of objects as well as the memories of disorder and loss. They also stared, as did Benjamin when unpacking his library, at "the floor covered with torn paper" (59). And like Benjamin, they anticipated replacing this confusion with a methodical arrangement. They were surrounded by a disorder which, in spite of their pains to categorize their specimens, ruined their efforts time after time.

The following pages explore not only the muddles and disappointments inherent in Victorian natural history collecting but also the consequences of that reality, particularly the resulting necessity to concentrate on skins rather than on live animals. In order to examine these circumstances, I have divided the chapter into three sections. After reviewing the Victorian enthusiasm for amassing and arranging exotic specimens, Part One, "Collecting and Classifying," enumerates the multiple disasters attending efforts to possess natural history specimens and illustrates how illusions of mastery disintegrated before collectors' eyes. Various catastrophes discouraged people from importing live specimens, even though they were frequently considered to be more desirable, and instead encouraged the collecting of skins. Part Two, "Skins," the most substantive segment of this chapter, concentrates on this preference for skin. The argument extends the discussion begun in the previous chapter concerning the significance of skin in representing identity. The chapter closes with an afterword, "In the Field," which examines the correspondence and notebooks of two avid Victorian collectors, the 13th Earl of Derby and his 22-year-old niece, Elizabeth Hornby. Their fascination with collecting live creatures as well as skins vividly illuminates Benjamin's understanding that the life of a collector rests uneasily "between the poles of disorder and order" (60). Their experiences help revise the well-rehearsed, yet, from my point of view, suspect convention of equating the Victorian practice of amassing natural history specimens with an imperialistic imperative. This section introduces new materials. In particular, the presence and diaries of Elizabeth Hornby.

---

[2] Sarah Whatmore in *Hybrid Geographies* is one of these critics who speak of this "colonial impulse to subject the world to systematic scientific account" as "one of the most authoritative actions in the exercise of government" (21).

[3] See Walter Benjamin's "Unpacking My Library" in *Illuminations*, 59–68.

Part One: Collecting and Classifying

Collecting natural history specimens was ubiquitous in Victorian Britain. The pursuit blurred typically rigid gender and class boundaries, for just about everyone participated: professionals, amateurs, scientists, travelers, hunters, naval officers, sailors, mill workers, shopkeepers, the elite and the poor, as well as both men and women.[4] Landed estates featured cabinets of curiosities, game trophies, cabinets of skins, as well as private menageries; middle-class parlors displayed aquaria, fern cases, butterfly cabinets, and, of course, stuffed birds; furthermore, working-class homes often boasted boxes of precious plants, insects, and fossils. As I have pointed out in the previous chapter, a number of working-class men in Manchester were "warm and devoted followers" of the "more popularly interesting branches of natural history." As Elizabeth Gaskell explained, they "knew the name and habitat of every plant within a day's walk from their dwellings" (39). Even aesthetes collected. In *The Picture of Dorian Gray*, while lounging on a "peach-coloured divan," Lady Narborough listens to the Duke of Monmouth ecstatically describing his "last Brazilian beetle," which he has just added to his collection (Wilde 163). He prizes the beetle for its color. Given these examples, Darwin's remark that he was surprised nobody had started collecting different-shaped biscuits is more apt than outrageous (Barber 17).

As part of this popular taste for natural history, people regularly visited museums, spent evenings at the microscope, and walked around zoos. The more serious among them kept notebooks of their findings, passed leisurely hours chipping geological samples from cliffs, and attended magic lantern lectures. Still others created societies or enrolled in a working man's college so as to increase their knowledge of geology, biology, and zoology.[5] The spaces of inquiry ranged from the professional dissecting rooms of the London Zoological Society to domestic interiors in which people "botanized in their kitchens and around their breakfast tables" (Schiebinger 164). Sharing a desire for a more perfect knowledge of the world around them as well as a growing curiosity,[6] they supported the momentum to pursue a systematic study of natural history. And acknowledging Britain's far-ranging power, they concurred with the Zoological Society of London's belief, expressed in an 1846 report, that "there is no community in the world possessing

---

4 Jane Camerini in her article "Remains of the Day: Early Victorians in the Field" remarks that "In Victorian times, collecting and studying plants, animals, rocks, and fossils were enormously popular activities, pursued by people as varied as the culture itself" (355). John M. MacKenzie also remarks upon the cross-class involvement in the collecting of natural history specimens. See his *Imperialism and the Natural World*. See also E.P. Thompson's *The Making of the English Working Class*.

5 For a lengthy discussion of the Victorians' enthusiasm for natural history, see Lynn Barber's *The Heyday of Natural History 1820–1870*.

6 See Katie Whitaker's "The Culture of Curiosity" in *Cultures of Natural History*, 75–90.

such manifold advantages for the study and critical investigation of the Animal Kingdom as our own" (Reeve 4).

Historians who reflect upon this interest refer to advances in technology, which allowed individuals to pursue their passion for collecting more easily. The development of railroads, for instance, offered people access to places, close to home, where they might gather various flora and fauna. In addition, expanding postal services permitted the sending and exchanging of specimens. Primarily, though, these historians concentrate on Britain's ability to roam the globe. Natural history collections testified to the reach of British commerce, which not only developed new routes to foreign lands but also peopled these distant places with entrepreneurs, civilians, commissioners, governors, and military personnel, many of whom already had a passion for collecting. Ever increasing commercial and administrative networks allowed these individuals to ship their spoils home or act as agents for those who were either intent upon discovering new species or in the business of importing and selling exotic creatures. In particular, the Royal Navy accommodated fieldwork and the acquisition of specimens.[7] As others have recognized, the work of naturalists, such as Darwin, Wallace, and Henry Walter Bates, was framed by the colonial experience. Their research in uncharted landscapes was incidental to a ship's official charge either to map the shores of a country or to investigate the area's commercial and strategic potential.[8]

Because the Victorian interest in natural history was intimately incorporated with the colonial enterprise, recent commentators have associated the business of collecting with Britain's imperialistic imperative to possess and dominate. As I mentioned at the beginning of the chapter, these critics exclusively regard the importing, classifying, and enclosing of exotic specimens (in zoos, museums, glass cases, cabinets, or drawers) as expressions of a wish to subject the foreign to British power. Kohlstedt believes, for instance, that the "collecting of animals gave people a sense of superiority and control" (4). She is among many who regard this activity solely as part of a desire to tame the alien other—a claim which, although at times accurate, is too all-encompassing. It leaves no room for the messiness of collecting and, hence, the loss of control and the immediate disorder. Some take the reductive parallel between collecting and imperialistic power even further, and equate the impulse to place these collections in a correct taxonomical order with Britain's campaign to claim command over its external regions. Lynn Barber, Janet Browne, Jane Camerini, Barbara T. Gates, John MacKenzie, Roy

---

[7] Jane Camerini's essay "Remains of the Day: Early Victorians in the Field" discusses how indebted natural historians were to the Royal Navy and how their fieldwork was situated in the colonial network.

[8] See Janet Browne's chapter "Biogeography and Empire" in *Cultures of Natural History* for a thorough sense of how natural history studies and empire are connected.

Macleod, and Harriet Ritvo are among many who see this correspondence. In "Zoological Nomenclature and the Empire of Victorian Science," Ritvo, for example, adamantly equates the conquest or possession of exotic territories with the desire to institutionalize and classify a region's natural resources. She claims that the British were reducing "these exotic areas to intellectual order" (334, 336).

This "order," however, was not easily to be had. To begin with, Victorian collectors and scientists interested in taxonomy could never rest contented within their classification systems, for no sooner was a species identified, than another would emerge to complicate the arrangement; moreover, debates ensued concerning the merits of rival systems. What had seemed authoritative and certain was frequently overturned. As Ritvo acknowledges, "over a thousand new genera were being named each year … with the species count correspondingly higher" ("Zoological Nomenclature" 336). A July 9, 1850 letter from John Edward Gray to the 13th Earl of Derby, acknowledging the reception of a skin, reflects this continuously shifting taxonomy: "Accept my thanks for your kindness in sending the skin & horn of the Barasingha [a swamp deer from North and Central India]. It is a most interesting addition to the Museum Collection & fills up an important blank" (National Museums Liverpool).

Because of such proliferation, collectors never fully experienced what Ritvo refers to as the "conquest of nature" ("Zoological Nomenclature" 334). Most certainly they failed to achieve the tight orderliness that one cartoonist in an 1896 issue of the German *Jugend* hyperbolically imagined might happen if European efficiency came to Africa and replaced the chaotic irregularity of the jungle with regimented lines of giraffes, elephants, lions, monkeys, crocodiles, and turtles. In the drawing, the animals, which had previously cavorted in a cluttered landscape, stand at attention while a colonial military officer inspects their ranks. No such fixed order was possible for nineteenth-century naturalists who increasingly realized that as fast as one "gap" was filled, "new yawning chasms appeared" (Barber 65). "Specimens not yet included in the Collection at Regent's Park," in a June 19, 1869 issue of *Punch*, spoofs this dilemma by mockingly depicting yet to be discovered species, such as "The Aquarious Lawn-Bird"—a showerhead that sprays water from the end of its beak—and "The Long-Sighted Hornbill," an African bird which sports binoculars for eyes (56: 258). Similarly, Edward Lear, the natural history illustrator and nonsense writer, as well as friend of the 13th Earl of Derby, as if acknowledging the dilemma, occasionally humorously abandoned the scientific labels to create his *Nonsense Botany* ("Manypeeplia Upsidownia") and his nonsensical sketches of birds, which he amusingly identified as "The Dark Blue Bird" or "The Orange-Coloured Bird," as if throwing up his hands at the impossibility of naming and classifying specimens.

For the Victorians, confrontations with all sorts of obstacles compromised dreams of dominion. Indeed, imperialistic dreams of subjecting the foreign, the

Figure 2.1 Cartoon from *Jugend*
Source: Jugend 1896.

alien other, through classification and possession were often transformed into nightmares, even after exotic creatures had been "tamed" and enclosed within British zoos and museums. As if aware of this reality, "The Keeper's Nightmare," in *Punch*'s 1871 "Almanack," humorously depicts what might happen if "all the animals [broke] loose and [swapped] heads": a giraffe's head joins a hippo's body; an owl's attaches itself to an elephant's torso. Chaos reigns. Gleefully asserting their own way and creating a senseless muddle, these liberated creatures usurp imperial authority. They make nonsense of the classification systems so carefully delineated within scientific institutions.

THE KEEPER'S NIGHTMARE.

ONE OF THE OFFICIALS AT THE ZOOLOGICAL GARDENS HAS A BEWILDERING NIGHTMARE. HE DREAMS THAT ALL THE ANIMALS HAVE BROKEN LOOSE AND SWAPPED HEADS, AND HE DOESN'T KNOW "WHICH TO FEED WITH WHAT."

Figure 2.2 “Keeper’s Nightmare”
Source: Punch 1871.

Disasters and Disorder

As the above examples suggest, a critical perspective which equates the imperialistic rage for dominion with the collecting and ordering of natural history specimens begs correction, for it does not necessarily correspond to the actual experiences recorded by Victorian naturalists. Reading through their narratives and reports, one learns that in spite of all their efforts correctly to label their spoils by recording the data of capture, their attempts were fraught with disasters, which wrecked their desire for orderliness. Individuals might, as the established taxidermist Rowland Ward demanded, mark skins for absolute identification by puncturing the owner's signature on the inside of the hide (*The Sportsman's Handbook* 18) or, as Montagu Browne insisted, note the data of the specimen's capture (*Practical Taxidermy* 117), but control inevitably slipped away. John Gould, the famous bird collector and illustrator, for instance, despaired when he discovered that the contents of one of his carefully labeled crates had arrived badly mildewed; moreover the bottles holding the specimens had not been properly secured, so the preserving spirits had escaped and "caused the small boxes to fall to pieces forming an heterogeneous mixture of nests, eggs, broken and sound, tops, bottoms, and sides of boxes, plants, etc." (Tree 111).

Popular as well as more serious adventure books were replete with these threatening and difficult moments, many of which reminded the reader that part of the allure of traveling to these faraway places was in actually engaging the ensuing chaos and disorder. Adventure is not possible, after all, unless one steps out of the comfort zone. Among these narratives are Roualeyn Gordon Cumming's *Five Years of a Hunter's Life* (1851), Paul B. Du Chaillu's *Explorations and Adventures in Equatorial Africa* (1861), R.M. Ballantyne's adventure novel *The Gorilla Hunter* (1876)—based upon Du Chaillu's book—and P.T. Barnum's *Lion Jack: A Story of Wild Men and the Capturing of Wild Beasts* (1877).[9] Entrancing their readers with stories of one catastrophe after another, they recounted "thrilling adventures which attended the capturing of savage beasts ... in order that they and their habits may be studied by civilized people" (Barnum iii). For instance, even though Du Chaillu brought home "2000 birds, of which more than 60 are new species, and ... killed upwards of 1000 quadrupeds" (viii), this collector for

[9] Roualeyn George Gordon Cumming (1820–1866) was a Scottish traveler and sportsman who collected natural history trophies and specimens. His collection of trophies was displayed at the 1851 Great Exhibition; Paul B. Du Chaillu (1831–1903) was a French-American explorer and anthropologist who became famous in the 1880s for confirming the existence of gorillas; R.M. Ballantyne (1825–1894), born in Edinburgh, wrote a series of adventure stories for the young. *The Coral Island* (1851) is the best known of these, for it is said to have occasioned R.L. Stevenson's *Treasure Island. The Gorilla Hunter* is one of these adventure books. P.T. Barnum (1810–1891), an American showman, is best remembered for the founding of the circus that became the Ringling Bros. and Barnum & Bailey. His *Lion Jack* is based on the capturing of beasts for his menageries.

the London Zoological Society and the 13th Earl of Derby told tales of severe disappointments and trying moments when the management or possession of a specimen were not to be had. Like many other collectors, Du Chaillu might have wished to replicate Hanno's ancient voyage to the African Continent, and upon his return to London triumphantly hang the gorilla skins he had amassed in the British Museum, perhaps the equivalent of Juno's temple in Carthage, but damage and disorder periodically attended the attempt.[10]

Imperialistic control over foreign territory was also hardly visible when, in the 1850s, Richard Owen of the London Zoological Society opened a crate containing a young male gorilla, sent by Du Chaillu to him from the interior of Gabon. Upon lifting the lid, Owen discovered a decomposed specimen—no antiseptic had been applied. Large patches of the cuticle with the hair had become detached. Much to Owen's horror, in this condition the young gorilla resembled "the naked boy of Man."[11] The drawings of this specimen are sad to behold, so too is the 1858 photographic portrait of A.D. Bartlett (of the London Zoological Society Gardens) with "his first preserved gorilla" (Vevers 66). These disasters were a far cry from Ballantyne's fictional and idealized description of a gorilla hunter unpacking his crates filled with perfectly conserved specimens: "All our packages and boxes of specimens arrived safely at the coast; and when unpacked for examination, and displayed in the large school-room of the station, the gorillas, and other rare and wonderful animals, besides curious plants, altogether formed a magnificent collection, the like of which has not yet been seen in Great Britain—and probably never will be" (*The Gorilla Hunter* 421).

As the above illustrations suggest, chaos (death and disorder) was an ever-present companion. The physical task of accumulating specimens in isolated areas was often dangerous. (One is reminded of Mary Kingsley, who, on her journey to collect fish in West Africa, fell into a rhinoceros trap and was saved from impalement by her thick woolen skirt.) And, of course, there were numerous instances of naturalists and their assistants getting killed. Moreover, according to various narratives, vultures swooped down and consumed specimens waiting to be carried to ports, hummingbirds died from traveling over bad roads, and live giraffes and hippos struggled for their survival across deserts on the backs of camels.[12] For example, Lovell Reeve, a Fellow of the London Zoological Society,

---

10 For Paul B. Du Chaillu's recollection of Pliny's account of this Carthaginian navigator's experience with gorillas, see *Explorations and Adventures in Equatorial Africa*, 343–4. According to Pliny's account, Hanno returned to Carthage from his voyage along the Western coast of Africa with the skins of gorillas he had captured and flayed. Hanno hung these hides as an offering in the temple of Juno.

11 See "Description of Young Gorilla" in Richard Owen's *Memoir on the Gorilla*, 5.

12 In P.T. Barnum's *Lion Jack*, Jack asks the Doctor: "'Haven't they [young hippos] been caught before?' … 'O, yes … Hayenbeck's men have done it, and had to carry them across the deserts on camel-back, with a lot of goats along to furnish milk for them'" (142). Major H.G. Swayne in his account of his trip through Somaliland and Abyssinia records

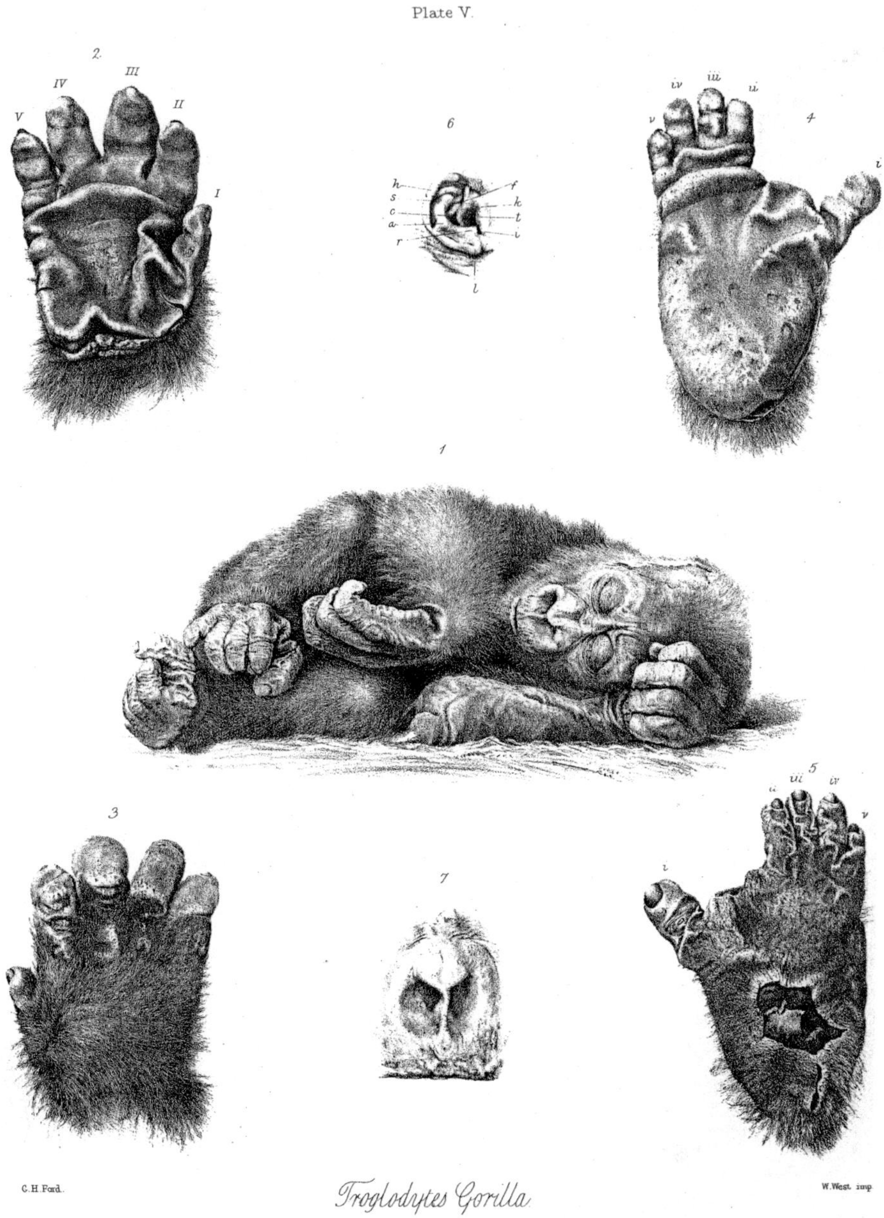

Figure 2.3 "Troglodytes Gorilla"
Source: Owen 1865.

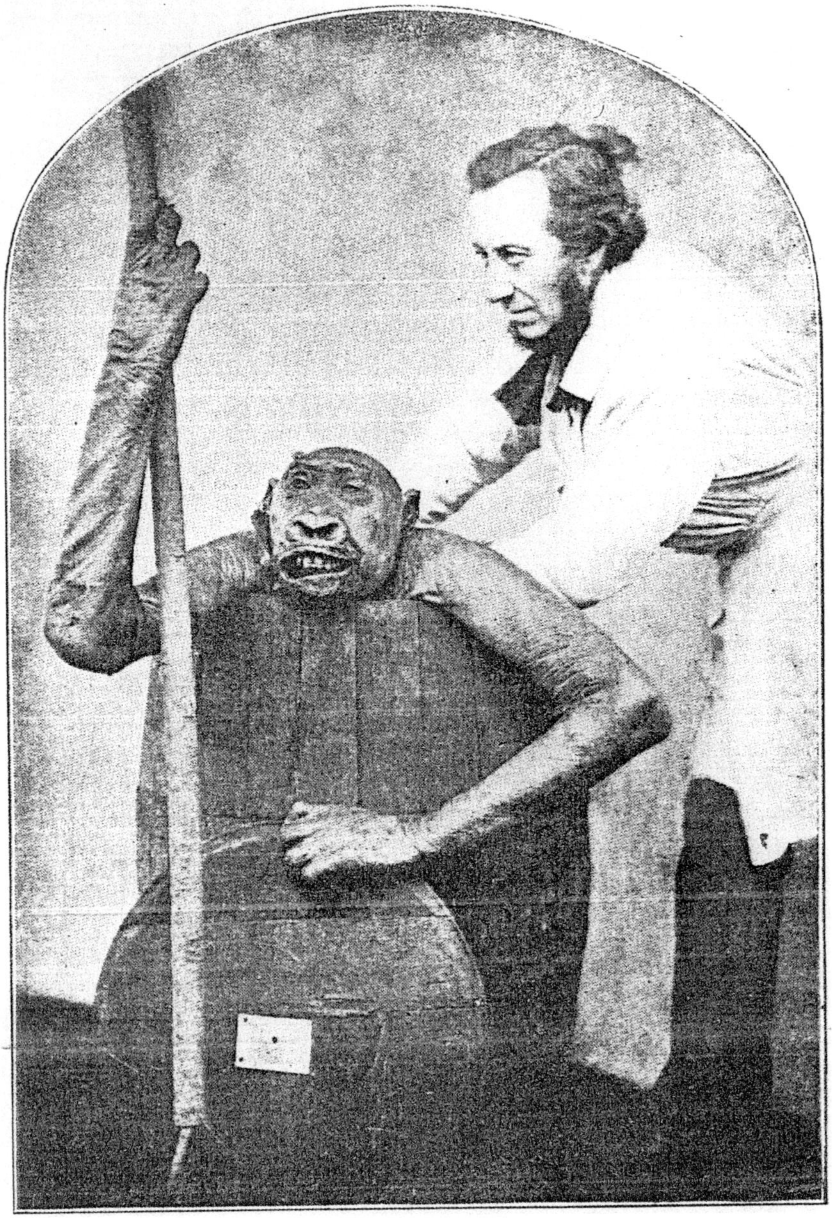

Figure 2.4 "A.D. Bartlett with his first preserved gorilla"
Source: Vevers 1858.

reported that "M. Thibaut quitted Cairo in April, 1834, with a firman from the Pacha, and on reaching the desert of Kordofan, in August, succeeded in capturing five young Giraffes. Four of these were killed by the severe weather, in the desert, on his return to Dongolah …" (7).

If the specimens survived the journey across a foreign land and reached the ship returning to England, there was the probability that they would not survive the voyage. Shipwrecks and fires on board destroyed collections (one only has to recall the disasters experienced by Sir Stamford Raffles in 1824 or by Alfred Russell Wallace in 1852).[13] High winds pitched crates into the sea; unruly animals were tossed into the ocean;[14] and live specimens died as a result of close confinement on board, improper diets, trauma, injury, illness, and cannibalism. On one occasion two elephants battled on board a ship; one pushed his rival into the sea. On another two large birds destroyed each other and pecked each other's eyes out. Sometimes, a voyage was longer than expected, so the sailors consumed the captured animals and birds in order to survive. For these reasons and more, collectors insured their finds and often sent, as did Wallace, duplicates on other vessels home.

Letters, diaries, and brochures expressed relief when a specimen arrived safely and not near death. D.W. Mitchell's 1852 guide to the London Zoological Society Gardens proudly records, for instance, that on May 25, 1850 the first hippopotamus "which had breathed on English soil since the Deluge" landed successfully at the quay at Southampton (54). In contrast to this, however, many other accounts document the all-too-frequent fact that a deer, a black-necked swan, or a giraffe had died before the vessel docked. In 1852, the Secretary of the London Zoological Society officially noted that "Sir James Brooke … shipped no less than five living specimens [orangutans] at his own cost for the Menagerie." He added, "It is greatly to be regretted that the object which he had in view in making this magnificent contribution was frustrated by the accident of the voyage, the last animals having died within sight of the English shore" (Mitchell 28–9).

---

a time when he shot two lions, and after laying the two carcasses side by side on the back of a camel, sent them back to his camp (104). There are many astonishing stories of the transportation of captured giraffes. The giraffe (or "Cameleopard" as it was once called) which belonged to George IV was, for instance, transported 1,200 miles across the desert and placed on the backs of camels (Huish 30). This giraffe expired in 1829 and was skinned and stuffed, at a cost of £148.10 by Messrs. Gould and Tomkins of the Zoological Society (Sauer 13). For an almost incredible account of an early journey of a giraffe to China on the back of a camel, see Sally K. Church's article "The Giraffe of Bengal: A Medieval Encounter in Ming China."

[13] The vessel in which Sir Stamford Raffles embarked for England in 1824 caught fire while only 50 miles from Sumatra. The passengers and crew escaped, but Sir Stamford's natural history collection and living animals were burnt. In Alfred Russell Wallace's case, his valuable collection together with his books and instruments went down when his Brig, the *Helen*, burnt at sea.

[14] In 1814 a hippo, which had been causing trouble on board ship, was tossed into the sea.

The narrative of loss was familiar. In the mid-1860s, when Clarence Bartlett, the London Zoological Gardens' Assistant Superintendent, went out to Surinam to take charge of an American manatee, he ultimately failed: "unfortunately the animal died a few hours before the vessel arrived at Southampton" (Scherren, *The Zoological Society of London* 138). Equally unsuccessful was the attempt to send one home from Puerto Rico. Bureaucracy could also cause problems. When, for instance, Charles Waterton was returning with some "curious" birds and eggs from Guiana, much to his disappointment he and his treasures were held up at customs in Liverpool; his hopes of rearing a new species of bird in England were dashed (*Wanderings in South America* 252). The docks themselves could also be disorderly and chaotic. Landing on British soil did not necessarily guarantee a neatly arranged environment. Specimens often found their way to jumbled curiosity shops jam-packed with the spoils and fragments from "far-distant and little-known climes" brought home by the sailors. Frank Buckland, a passionate Victorian naturalist, describes these places: teeth of a whale, the "monster of the deep," lay "mixed with the whalebone from the capacious mouth." From the ceiling hung cages full of tropical birds; below could be heard the "harsh and unearthly noise" of a tailless, green-coated African monkey suspended from "an old parrot's cage" (*Curiosities of Natural History* 291–2).

Even if a live specimen arrived safe and sound, once enclosed within a cage it would periodically escape and sometimes remind people that the subjection of the foreign was not always feasible. At times people literally ran to save their own skins from a wayward lion, or from a black bear which had climbed up a chain and escaped from a zoo's bear pit. In the 1840s, a fugitive kangaroo hopped through

Figure 2.5 "Escaped Kangaroo at Regent's Park"
Source: Cruikshank, ca. 1840.

Regent's Park and alarmed everyone; and in 1851 two polar bears escaped from a Liverpool menagerie. They were eventually rounded up in a cemetery. In 1867 a polar bear from the London zoo scaled the wall and climbed out of the top of its cage; in the 1870s a sea lion escaped over a 3-foot-high iron railing; and in 1893 a fugitive boa constrictor caused concern at the Belle Vue Gardens. Moreover, an elephant trained to give children rides suddenly refused to follow commands and raced through the grounds with the frightened children on his back.[15] Being within the civilized spaces of an English zoo did not ensure domestication. A.D. Bartlett records, for instance, that a rhinoceros repeatedly and "savagely" attacked the bars of its cage; a chimpanzee also flew at a keeper's neck (*Wild Animals in Captivity* 64, 134).

Part Two: Skins

Why Skin?

Because of all these uncertainties and difficulties, people generally agreed that the collecting of skins, rather than live specimens, was a relatively manageable and certainly a more orderly and reasonable endeavor.[16] Compared to a struggling and all-too-mortal beast, skins were obviously more durable and portable; they could be labeled or stored comparatively easily. Moreover, with little trouble, they could be reshaped into trophies to display the metropole's power (think of the tiger skins draped over a divan or mounted on a wall in an English estate). For those not necessarily intent upon such an imperialistic agendum, these skins were also thought to be of invaluable assistance. Scientists as well as collectors interested in classifying or identifying a particular species, relied upon skins to help them compose a more inclusive portrait of the natural world. Skins, once removed from temporality as well as origin, could be compared in order to allow scientists to create a seemingly inclusive picture of the globe.[17] Gathered and displayed in museum cases, for many they created a fictional portrait of empire.

---

15 As I write this paragraph in the spring of 2012, news has just come over the radio that the gorilla from the Buffalo Zoo has escaped from his cage and nipped his keeper. (The keeper found safety by entering the cage of a female gorilla and baby.) All has ended well, however. After being tranquilized, the gorilla is back in his area; the keeper is doing well. The *Buffalo News* featured photographs of the public fleeing and standing in safe areas.

16 Anne Larsen, in "Equipment for the Field," remarks that specimens "were manageable pieces of the natural world that could be bought, sold, exchanged, transported, catalogued, displayed and consulted by many people" (358).

17 Susan Stewart in *On Longing: Narratives of the Miniature, the Gigantic, the Souvenir, the Collection*, remarks: "the museum of natural history allows nature to exist 'all at once' in a way in which it could not otherwise exist" (162).

Even those individuals primarily interested in gathering live specimens acknowledged that skins significantly contributed to the study of the natural world and were integral to creating a more inclusive picture. For instance, in 1847, the Secretary of the London Zoological Society admitted that from "a purely scientific point of view I apprehend that a collection of skins is infinitely more available and is certainly far more capable of inexpensive conservation" (920/DER/13 National Museums Liverpool). His words echoed those uttered earlier by Mrs. R. Lee, who accompanied her former husband, T. Edward Bowdich, on excursions to Africa and assisted him in the preparing of skins. In the 1830s, reaffirming the cautionary words of Bowdich's 1820 taxidermy manual, she acknowledged that given the dangers in shipping live specimens, "We must content ourselves by bringing the skin, head, and feet of the large animal we have killed in too distant a place to transport them entire" (Lee 152).[18]

Skinning, as a consequence, was pervasive. Almost suggestive of slave traffic, which agents and hunters such as Cumming and Du Chaillu witnessed while tramping among slave traders in Western Africa or crossing the oceans, avid collectors dealt in skins and instructed assistants to bring home as many hides and pelts as possible.[19] Gould, for example, insisted that his assistant, John Gilbert, "Collect skins of *all* kinds of kangaroos in New South Wales and other mammalia" (Tree 134). Big game hunters as well as government officials, who traveled to or resided in colonial outposts, were encouraged to collect and prepare skins. In 1840, William Swainson proclaimed: "Gentleman residents abroad, particularly in India, have it very much in their power to benefit the museums of this country, by acquiring a sufficient knowledge of taxidermy to enable them to preserve the skin of animals killed in the chase, since the public and private collections in Britain are very deficient in many of these species." He instructed, "The skins may be removed and slightly stuffed" (3). The result was, of course, that museums and private collections were overflowing with these spoils. "Cabinet skins" (drawers, lined with cotton wool, of labeled skins belonging to thousands of birds, mammals, and reptiles) as well as those skins stuffed by taxidermists (these are the specimens that museums now frequently either hide or apologize for) contributed to the overcrowding in these spaces.[20]

---

[18] In his history of taxidermy, P.A. Morris points out that Mrs. Lee's taxidermy manual is simply a republication of the one written by her husband, Thomas Edward Bowdich. Over 20 years after his death in 1824, she took it upon herself to confirm his reputation in the field. However, it is clear that she herself was conversant with the art. She not only had assisted her husband but also had observed and taken notes on the taxidermy techniques used by Charles Waterton (336–7).

[19] See Paul B. Du Chaillu. *Explorations & Adventures in Equatorial Africa*, 185.

[20] On a recent visit to the Museum of Natural History in London, I was interested to note that the curator had placed an apologetic sign next to every stuffed animal or bird. The label explained that each of these specimens was inherited from the Victorian age. Contemporary distaste for stuffed creatures also came to the fore when protestors recently picketed the Berlin Zoo to stop officials from handing a deceased and much beloved polar

From the point of view of most Victorian collectors, however, skin was by no means a poor or inadequate substitute for the living creature. Contrary to what Mrs. Lee believed, skin was *not* something with which merely to be "content" (152). Given its importance in establishing distinctiveness, skin was thought, instead, to be invaluable and necessary. It was, as we have seen in the previous chapter, essential to establishing identity and creating portraits of empire.

As the first chapter suggests, skin was the place where identity was assigned. It was the primary medium through which people differentiated species and ordered the natural world. A detached skin provided a surface which collectors treated as a text to be read and interpreted, and to which interested parties assigned meaning and nomenclature. Assembled with other skins, it helped scientists create a more perfect and systematic record of specimens from all over the world; it also allowed people, for the first time, more accurately to describe rare mammals and birds from remote places. As we have seen, Owen depended upon shipments of gorilla skins from Gabon to increase his knowledge of the species. Others too relied upon skins sent to them from distant lands so they could learn about creatures they had not known of before. In *Life Among Wild Beasts*, A.D. Bartlett explained that it was only after examining the skin of a bird that he realized "at once" that it (a guinea fowl) was "unlike anything I had seen" (174). And after looking at a group of skins, he was able to determine that the spotted emus in his collection were "smaller than the common species" (210). As might be expected, Gould's correspondence is replete with references to the link between skin and identification. He recognized, for example, a spectacled cayman because of the markings round its eyes (Sauer 137).

Victorian collectors regarded skin as a "worn identity" (Benthien 161). For them, skin was comparable to a garment, a dress; it was a covering that represented the subject's uniqueness and signified the creature's character and position in the world.[21] No wonder, then, that Roualeyn Gordon Cumming, one of the most prolific mid-century hunters, wrote about "disrobing" an African lion he had just shot. In one account of a lion hunt, he explained, "I considered myself extremely fortunate in having secured so noble a specimen of the lion with so little danger, and I at once set men to work to *unrobe* him, which they were not long in accomplishing" (*Five Years of a Hunter's Life,* 1: 265 [italics mine]). Given this metaphor, it is not surprising that Cumming was intrigued by the fact that the "wild Bushmen," when

bear, "Knut," to a taxidermist so he could be put on display. "Stop Knut being stuffed" they chanted. See *The Independent*, Wednesday, April 13, 2011: 26.

21 Claudia Benthien's discussion of skin as clothes is in chapter 8 of *Skin: On the Cultural Border Between Self and the World.* She is aware that for many human skin possesses the character of a "garment" and that for them "ethnic identity is merely a worn identity" (161).

hunting ostriches, donned the skin of the bird and wore it like a robe so as to draw closer to their target.[22]

Others also described an animal hide as if referring to a garment or to a piece of cloth. For instance, in order to distinguish between an Indian and African rhinoceros Wilhelm Kuhnert compared the distinct folds of the textile/material/skin covering the two animals. In his turn-of-the-century *Animal Portraiture*, he remarked: "A satisfying and easily recognized distinction between the rhinoceroses of south-eastern Asia and the two African species is afforded by the circumstance that while on the former the hide is thrown into a number of deep folds dividing it into separate areas, on the latter these folds are more or less lacking, so that the skin is comparatively smooth like that of a pig." As if differentiating one piece of material from another, he concluded: "The shape of the skin-folds serves to distinguish the species" (60).

Skin was also considered valuable because it registered the fact that a person had actually set eyes upon the subject. It was a confirmation of seeing. As Steven Connor suggests, skin serves as "proof of our exposure to visibility itself" (51). It was, however, more valuable than a souvenir. Skin was no substitute, as is a souvenir, for the real thing. It was not simply a trace of the genuine being; it *was* the real thing. As the authentic outer covering of the entire body, skin, more or less, retained the creature's proportions, size, color, and texture. Unlike the souvenir, it was neither miniaturized nor reproduced. Furthermore, for the person who had bagged the skin it carried with it the memory of the original context or remote experience. In this respect, skin was not merely a trophy but also an intimate and significant piece of tangible evidence. For these reasons, if it were impossible to transport a live specimen, or if a captured creature died on the return voyage, the desired practice was to remove the skin and display it for others to see.[23] Berg, in Barnum's 1877 account of capturing wild beasts, knows he cannot bring back the hyenas he has just encountered, so he shouts out to his companions: "We must have the skins of those fellows, if we can" (108). And after a giraffe died on board the *Lady McNaughton* in 1831, the decision was immediately made to reserve the skin of it (Scherren, *The London Zoological Society* 61).

---

22 Roualeyn Gordon Cumming in his *Five Years' Adventure in the Far Interior of South Africa* writes: "A favourite method adopted by the wild Bushman for approaching the ostrich and other varieties of game is to clothe himself in the skin of one of those birds, in which, taking advantage of the wind, he stalks about the plain, cunningly imitating the gait and motions of the ostrich until within range …" (75). This practice was quite common. For instance, George Catlin's mid-nineteenth-century painting of a Buffalo hunt depicts him and his guide, hidden under white wolf skins (including the head), stealthily approaching a herd (Smithsonian American Art Museum).

23 Skins were not just shown in the halls of the wealthy who were eager to show their authority but also continually put on display to illustrate natural history lectures. One should also not overlook Gould's 1851 display of 1,500 stuffed hummingbird skins shown at both the Great Exhibition and at the London Zoological Gardens. The exhibit attracted 75,000 visitors.

As a visible envelope of a self or a body, skin was also deemed important because it metonymically stood for the entire being; it displaced the part for the whole.[24] In her study of skin, Claudia Benthien observes that skin often functions "as a stand-in for a 'person,' 'spirit,' 'body,' or a 'life'" (13); she proposes that skin "metonymically [represents] the whole human being" (17). Connor also remarks on this capability when he suggests that skin is "an entire environment," which has the "capacity to resume and summarize the whole body" (29).

DISCOMFITURE OF OLD MR. J—N—S, WHO, ON VISITING A PRIVATE COLLECTION, MISTAKES "PETER," THE GREAT HORNED OWL, FOR A STUFFED CAT.

Figure 2.6 "Discomfiture of Old Mr. J—N—S"
Source: Punch 1856.

[24] Susan Stewart recognizes that part of a collection's gesture is its "metonymic displacement of part for whole, item for context" (162).

Many Victorians would have concurred with these observations because for them skin had the capability of appropriating the totality of the subject's physical being and making what is absent present. They believed that by concentrating being on the body's outermost edges, skin had the power to summon the whole creature. As a result, they not only collected the skins of animals and birds they were unable to bring back from distant lands but also retained the skins of those creatures that had survived the voyage but had subsequently succumbed to accident or illness. The latter was at work when A.D. Bartlett was called upon to skin an elephant that had died in a traveling menagerie. Unable to convey the dead elephant by train to London, he accepted an invitation to set up a tent in the large garden of a gentleman, living in Camden Town, and remove the skin there. (The understanding was that the gentleman would receive the cut up flesh for his dogs.) The task took him and his assistant several days, during which time a "crowd of two or three hundred" assembled to watch. The skin was later mounted and displayed at the Crystal Palace (*Wild Animals in Captivity* 55). Similarly, when an ostrich died from overeating at the Royal Menagerie at Windsor, its "skin was accordingly preserved" (Vevers 110). This procedure was thought not only to represent but also in a tangential way to revivify the creature's being. Indeed through the techniques of taxidermy, a skin, the visible envelope of the body, could be stuffed and shaped so that an animal or a bird could once more pose in a lifelike manner or perch as if alive. Resisting the ravages of time, a preserved skin now covered an almost imperishable body. A cartoon in an April 12, 1856 issue of *Punch* humorously illustrates the fine line between life and death in taxidermy. Old Mr. J—N—S, when visiting a private collection, "mistakes 'Peter,' the [very much alive] Great Horned Owl, for a Stuffed Cat," and suffers the consequences (30: 144). He, unfortunately, has not been able to distinguish between the stuffed and the live specimen.

Accident, Decay, and Contamination

Though the collecting of skins was thought to be more feasible and reliable than the assembling of live creatures, there were, however, disrupting circumstances that made this alternative also susceptible to accident, decay, and contamination. For all their advantages, skins were not entirely exempt from the kinds of problems facing the amassing and delivering of live specimens. Order and authority were almost as elusive.

Wallace's description of his working/dissecting room where he would skin his specimens while in the Malay Archipelago captures the less than ideal milieu available to most collectors in foreign lands. The space's confusion reflects the larger disorderly context. It would have been difficult to exert control and power here:

> One small room had to serve for eating, sleeping, and working, for storehouse and dissecting-room; in it were no shelves, cupboard, chairs, or tables, ants

> swarmed in every part of it, and dogs and cats, and fowls entered it at pleasure. Besides this, it was the parlor and reception room of my host, and I was obliged to consult his convenience and that of the numerous guests who visited us. My principal piece of furniture was a box, which served me as a dining-table, a seat while skinning birds, and as the receptacle of the birds, when skinned and dried. (*The Malay Archipelago* 169)

Even in a laboratory or a workroom superior to what Wallace depicts, the very act of removing the skin from a creature was continuously subject to the vagaries of accident or disaster. A slip of the hand while separating the skin from the body caused the knife to cut too deeply. The consequent blood and mucus, as well as other bodily fluids, damaged the feathers or spoiled the fur. Also, sometimes in the process of skinning a large game trophy, the worktable or cable broke, and the animal, supported or hoisted by such a mechanism, dropped and injured a person. In 1847 while A.D. Bartlett was skinning an elephant in the dissecting rooms of the London Zoological Society, the tackle holding up the creature's body malfunctioned; the elephant fell and almost crushed him (see Blunt 164).

Skinning was not a simple art. As the above examples suggest, it could be tedious and difficult.[25] Out in the field the "initial priority" was "that the bird or animal be correctly skinned" but this directive was not always possible,

Figure 2.7 "Skinning a Tiger"
Source: Ward 1894.

[25] To read descriptions of how to skin an animal or bird, see William Swainson's *Taxidermy, a Bibliography, and Biography* (1840); Mrs. R. Lee's *Taxidermy: or the Art of Collecting, Preparing, and Mounting* (1843); Samuel Wood's *The British Bird Preserver* (1877); and Montagu Browne's *Practical Taxidermy: A Manual of Instruction to the Amateur in Collecting, Preserving, and Setting up Natural History Specimens of all Kinds* (1876).

especially when naturalists were away from home and moving across foreign plains or through dense forests (Frost 12). Unless collectors brought along their own assistants, they relied upon people who, according to their travel narratives (many of which reflect racist assumptions[26]), were, at times, neither trustworthy nor skilled and who botched the skinning. On one occasion, after killing "a rare and unknown bird," Du Chaillu was unable to preserve this treasure because his African assistants had "slyly" eaten its flesh and, consequently, ruined the skin (23). (His African cook also attempted to spoil him by adding two tablespoonfuls of arsenic to his soup! [245])

Moreover, at times naturalists found themselves subject to indigenous methods, which, from the collector's perspective, could be disastrous and acted as a reminder that colonial authority did not extend to everything. Wallace was among many who were sporadically dependent upon the local population for acquiring the skins of certain species. On one occasion, Wallace wanted to send back the skins of the Birds of Paradise, a species just introduced into British collections. To accomplish this, he relied upon the natives to achieve this goal. However, their preserving techniques damaged the skins. Recalling one painfully disappointing incident, Wallace explains that he was presented with one, the "beautiful head" of which was "all defiled by pitch from a damman torch; another had been so long dead that the stomach was turning green." Wallace consoled himself by noting that "Luckily, however, the skin and plumage of these birds is so firm and strong, that they bear washing and cleaning better than almost any other sort; and I was generally able to clean them so well that they did not perceptibly differ from those I had shot myself" (*The Malay Archipelago* 536). Similarly, Gould despaired when he received Bird of Paradise skins which had been "prepared in the traditional way" by the inhabitants of Papua, New Guinea (Tree 64). Sensitive to these sorts of impediments, a person writing on Gould's behalf insisted on finding a reliable person to shoot and skin the hummingbirds, the skins of which Gould was collecting for his volumes of *Humming-Birds*. After complaining about the fact that some hummingbirds had been shot with too large a bullet (hence ruining their skins), he insisted: "The great point is to have the specimen *properly* skinned. Mr. Gould says *properly* because herein lies the difficulty, a difficulty far greater than obtaining the birds." Impatient, he continued: "Surely some person able to do this may be found among the Medical officers of the islands or the young Surgeons of the Ships visiting it" (Tree 161).

[26] Phrases such as "monkey-faced races," "woolly-headed and thick lipped," give some sense of the racism. In a letter from the curator of Knowsley Hall, written March 17, 1842, Louis Fraser told Lord Derby, "I have a man at Cape Coast who promised to collect me many things for your Lordship, but there is no trusting to a Black Man's word I can say no more about it" (National Museums Liverpool).

Skin collectors were confronted by yet more troubles. By night, predators lurked about the camp in an attempt to snatch the removed skin.[27] And by day, specimens rotted in the heat of the sun; the skin dried out or shrank, as well as lost its color. Moreover, inevitably, from time to time, decay attended the shipping of skins. As a matter of simple fact, one taxidermist warned: "Rare birds are frequently received from foreign countries, the skins of which are in such a state of decay, that it is impossible to mount them by the ordinary processes … The only way in which they can be preserved, is to mount them feather by feather, which however is a very tedious method" (Brown 45). In the same manual offering "Instructions to Travellers," Captain Thomas Brown, hoping to prevent damage, also suggested: "Before skins of any kind are packed, they must be well shaken, and then rubbed over inside with oil of petroleum, or spirits of turpentine; these must be applied with a brush. They are then slightly packed with cotton, but just sufficient to prevent the inside of the skins from pressing on each other … The box in which they are packed must be pitched all over to prevent damp and air from reaching the inside" (128).

Foreign Contagion

Brown's insistence on applying turpentine to a skin reflects yet another obstinate reality: the ravages of the insect population. When shipping and displaying the skins of specimens, collectors had to protect their spoils from various destructive pests. More than any other disastrous element identified above, these tenacious, yet almost invisible, creatures undermined illusions of visible imperialistic control. They were the plague of all collectors. As agents of a cultural invasion, insects represented an imminent threat. Consequently, those dealing with imported skins treated these parasites as if attending to carriers of infection and disaster. Indeed, travelers' descriptions of ruin caused by insects remind one of what Priscilla Ward has recently identified as the "outbreak narrative" (*Contagious: Cultures, Carriers, and the Outbreak Narrative*), for they not only identified the source of the insect infestation (usually in foreign lands) but also followed the routes of transmission. By equating the ensuing virulent destruction with the spread of disease, their stories promoted the stigmatization of the foreign.

According to Victorian narratives of collectors' adventures abroad, larvae, eggs, and armies of ants silently assembled and waited for the right moment to infiltrate and devour a choice specimen. For instance, in spite of all of Du Chaillu's bravado as well as his seeming ability to dominate a continent as well as its wildlife, the threat of these beings made him apprehensive. If he did not take the proper precautions, he knew these pests would destroy the skins he had accumulated during his African tour. Annoyed and apprehensive, he wrote:

---

[27] Rowland Ward, in *The Sportsman's Handbook to Practical Collecting*, warns: "All skins should be well looked after at night, as during the darkness, animals of various kinds … frequently lurk about a camp" (27).

> I begin now to have so many animals on hand that I find I cannot go as far as the Nazareth. The risk of losing all my collection is too great; and the trouble of taking care of it is greater still. The ants—those little pests—are constantly on the look-out for prey; and it is impossible to leave a dead animal about for the shortest time without imminent risk of having it destroyed. Let the reader bear in mind that not only has the hunter-naturalist in these African backwoods to kill his game, which may occupy all day, but when he comes home tired he must immediately stuff it that it may not spoil, and then it must be suspended by strings from the rafters of the house to keep the ants off. The slightest carelessness may bring ruin upon his most cherished specimens; and I have more than once been reduced to the brink of despair by finding a choice bird or other animal in one night, and through one slight oversight in the preparation of the suspending cord, completely riddled and ruined by ants in the morning. (159)

Transposing these conditions into a fictional mode, Ballantyne's *The Gorilla Hunter* similarly describes the utter defeat of the collector who finds his "beautiful birds" eaten and ravaged by these creatures: "Alas! When I came to examine them [the bird skins] next morning, I found that … the white ant, had totally destroyed the greater part of them, and the few that were worth stuffing were very much damaged" (108–9). He begrudgingly acknowledges what taxidermists stressed: preserve the skin as soon as possible.

> Experience is a good, though sometimes severe, teacher. Never again did I, after then, put off stuffing of any valuable creature till the next day. I always stuffed it in the evening of the day on which it was killed, and thus, although this practice cost me many a sleepless night, I preserved, and ultimately brought home, many specimens of rare and beautiful birds and beasts, which would otherwise have been destroyed by the rapacious insects. (109)

With their ability to invade and destroy skin, overnight as well as over time, insects spread, as does a contagious germ (like Blake's invisible worm[28]), from one skin to another. As if reverting to the seventeenth-century practice of soaking letters sent from plague-ridden London in order eliminate the pestilent matter, Victorian collectors soaked the hide or feathers with turpentine and rubbed the inside of the skin with arsenical soap and burnt alum.[29] They knew too well that the skin envelope carried within it destructive foreign insects, which multiplied and spread to other trophies. Aware of how easily these alien insects could creep through the tiniest crack and hide within the smallest of the skin's recesses, taxidermists and museum officials warned that precautions were an absolute

28 See William Blake's "The Sick Rose" from *Songs of Innocence and Experience*.

29 Even letters from the capital were treated as if they were poisonous. Letters were variously scraped, heated, soaked, aired, and pressed flat to eliminate "pestilent matter" (http://ocp.hul.harvard.edu/contagion/concepts.html).

necessity. When Gould, then the Animal Preserver to the British Museum, testified before the 1836 Select Committee on the British Museum, he answered anxious questions concerning the condition of the Museum's specimens. In response to one question, he replied: "the specimens are not in the best state of preservation, they are not so good as I have seen at other collections, and they are not in that perfect state in which I should like to see them" (Sauer 127). A member of the Committee, Sir Philip Egerton, also questioned Gould specifically about moths in the storerooms and the reputed fact that "the best specimens are exposed to dust and insects" (Sauer 129). Yet another member asked him whether or not it would be possible to exhibit a stuffed camel and not have it suffer from moths (Sauer 133–4). This sort of anxiety was familiar beyond the hallowed walls of the British Museum. Indeed, it was still thriving in 1869 when *Punch* printed the cartoon "Sanitary Measures." The drawing shows three people painting a coat of varnish on a stuffed rhinoceros. The caption reads: "The Rhinoceros (British Museum) Has been Suffering from Parasites—Gover'ment Orders Him to be Varnished." As if the nation were warning against foreign contagion, a sign next to the animal warns: "Visitors are requested not to touch" (56: 140).

Storage areas where light could not reach the specimens were also a problem. To prevent an infestation, William Swainson prescribed that "Once or twice every year, each specimen should be carefully examined, and the fur and skin well saturated with spirits of wine and corrosive sublimate, in the proportion of three drachms to a pint" (*Taxidermy* 82). Disasters did, however, occur. Indeed, in 1883, the British Museum's Annual Report stated that of the 5,500 specimens (ironically, of insects) listed in an earlier catalogue, most had deteriorated to the extent that literally not one vestige of them remained.[30] Furthermore, thousands of unmounted skins sat rotting in the basement. Reports from the London Zoological Society also spoke of the disorder in the dark recesses allotted to the non-exhibited specimens. Officials stressed the danger to the entire Museum and its priceless contents. As if replicating the Great Fire of London (1666), which reputedly purged the city of the plague, the Museum periodically attempted to put an end to the contagion by making a bonfire of its rotting specimens. The stink was pungent. These conditions contributed to Darwin's informing John Henslow that he probably would not choose to send his specimens from his travels either to the London Zoological Society Museum or to the British Museum. They were too chaotic: "The Zoological Museum is nearly full, and upwards of a thousand specimens remain unmounted. I dare say the British Museum would receive them, but I cannot feel from all I hear, any great respect even for the present state of the establishment" (Blunt 136).[31]

---

[30] See Barber 161.

[31] In an effort to create more order in the British Museum, the authorities began selling off its duplicates "on a large scale" in 1849 (Scherren, *The Zoological Society* 99). A September 29, 1883 cartoon in *Punch*, entitled "A Bloodless Battue," makes fun of these duplicates. The caption reads: "In the Name of Humanity Lord Barndore Wrings the necks

Quarantine was yet another way of dealing with this foreign contagion. Taxidermists, museum officials, and collectors prescribed not only prophylactic measures but also isolation. They counseled that "The moment the feathers of a specimen appear discomposed, falling off" in a case, "the contents should immediately be examined, and the infested specimens removed." If done immediately, they believed, this measure would "frequently stop the progress of *contagion*" (Swainson, *Taxidermy* 85–6 [italics mine]).[32] As if resorting to the practice of isolating and barring the doors of those suffering from the plague, they also insisted that "Skins should be … invariably kept distinct from the collection, until the destruction of the insects is well ascertained … The seams or openings should be rendered air-tight that the fumes [of the pesticide] may not immediately escape" (Swainson, *Taxidermy* 85–6).

As the cartoon about the varnishing of the rhinoceros suggests, touching a potentially contagious object was also forbidden. Not touching was another form of quarantine. One taxidermist advised:

> In packing objects of natural history, attention is required to see that they are well preserved from the attack of insects; the skins must be well rubbed *inside* with the arsenical soap. Caution must, however, be exercised in the use of it; and in mounting the skin afterwards, it is necessary to avoid touching them with the hands as much as possible, as it often produces disagreeable effects. Those objects which are prepared with the preservative should be marked, so that the stuffer may avoid touching them too much. (Brown 128)

While amassing skins and trophies in South Africa, Cumming painfully experienced the effects of such contamination. Feeling cold after a long day of hunting, he fell asleep with one of the flayed skins over him. Upon awaking in the morning, he discovered that his own skin was "thickly tenanted by small transparent insects" which had "banqueted" upon his flesh and left his whole body "poisoned and inflamed" (*Fiver Years of a Hunter's Life*, 1: 358). The marks on his skin were proof of his having touched an alien other. His skin, the place of encounter, gave his transgression away.

---

of all his Pheasants, and having stocked his Preserves with Duplicate Specimens from the British Museum, invites his friends for a Day's Shooting" (85: 145). Scattered around his grounds are stuffed specimens of exotic birds.

32 Another example is: "Every precaution has been taken to secure this specimen against the attacks of moth; should any however have deposited their eggs on the specimens before being enclosed in the case they will show themselves within a few months … A guarantee is given that should this occur within twelve months of appended date the specimen being partly or totally destroyed, it will be repaired or replaced by a similar one where possible, or one of equal value" (Frost 12, quoting William Furman).

Conclusion

In spite of all their vagaries, skins were, however, the most intimate and tangible fragments of the natural world. Though sometimes thought by some, such as Philip Gosse, to be "blackened" and "shriveled,"[33] they still outlined the life that had once inhabited them. Transformed into the collection, they functioned as a metonym for identity; enclosed and arranged in cases, they offered an illusion of order. Settled among classifying labels dictated by what Gosse terms "the mystic cloud of Graeco-Latino-English phraseology," these skins reflected a sense of authority and certainty (Barber 41). As I have pointed out when talking about collections of exotic animals in zoos, the Victorian public would have glanced at these classified specimens and believed they were gazing at a framed and fixed portrait of empire: each was a miniature of an exotic place; assembled, they were a grand regimented portrayal of Britain's global reach. Today, however, when we stare at the remnants of these Victorian skins, we return them to the disturbing framework of time and history, and, thereby, to a context the nineteenth century tried hard to circumvent. Now no longer a frozen and secure moment, each specimen admits of fluctuation and decay. Loss and disaster have become their primary narratives. Looking at the skins of species, either extinct or threatened, today's public recognizes a receding natural world. (This is why many museums attach apologetic notes explaining that such collecting is no longer the institution's practice.) Because these skins acknowledge our environment's vulnerability, Chaos comes to the fore and compels many to avert their eyes from the skins' surface markings and colors. Scientists, in particular, turn away from the visible skin, and instead study what is invisible to the naked eye: they microscopically search within the skin's individual cells for recurring DNA patterns so that they may locate order and identity. In this way, they try to explain a world lying delicately balanced between order and disorder.

Afterword: In the Field

The passion for collecting skins as well as live specimens comes even more to life if one looks more closely at the experiences of individuals dedicated to this practice. Two avid Victorian collectors worth exploring are the 13th Earl of Derby and his young niece, Elizabeth Hornby. Their letters and diaries offer an immediate glimpse of the accompanying frustrations which from time to time threatened to upset their interest in obtaining specimens and classifying the natural world. They carried out their activities, sometimes uncomfortably, within what Benjamin identifies as the dialectic of order and disorder. The sheer curiosity and pleasure of these two, as well as their willingness to cope with all sorts of disasters, invite one to look beyond the much-rehearsed paradigm about collecting within the colonial context. Their attitudes toward their collecting displayed their sense of wonder and

---

33 See the quotation from Philip Gosse which opens this chapter.

curiosity—their fascination for the rare, the novel, and the outstanding. They were as much interested in provoking wonder as they were in creating order.[34]

The 13th Earl of Derby (1775–1851)

Edward Smith Stanley, later the 13th Earl of Derby, was known internationally for his impressive knowledge of zoology and botany.[35] From 1828 to 1834, he was President of the Linnaean Society, an organization devoted to natural history classification. He was also a founding member of the Zoological Society of London, and, from 1831 until his death, its President.[36] His estate, Knowsley Hall, situated a few miles from Liverpool, had its own well-stocked natural history museum as well as a 100-acre private menagerie. The latter was said to be the most extensive in Britain.[37] He was obviously a passionate collector. At the time of his death, over 20 agents in various parts of the globe (e.g. North America, Central America, South Africa, South America, India, Australia, New Zealand, and China) were under contract either to secure live creatures or purchase skins for his own collections as well as those curated by the London Zoological Society.

Anyone interested in the absorbing commerce of exchanging, perusing, pricing, acquiring, and shipping of specimens should examine the notebooks and letters exchanged between the Earl of Derby and his representatives.[38] They offer an extraordinarily detailed picture of the business. One learns innumerable telling particulars about the transporting of live animals. For example, on April 25, 1839, Rev. John Fry, his agent in South Africa, sent him word that he was about to ship a group of creatures. The list of foods to nourish them during the voyage is instructive:

> To morrow I ship the six ostriches, the male lion, the male Buffalo, 4 Secretaries, 3 Blue Cranes and a kettle of turtles all of which I trust you will safely receive. I ship them with 20 Goats, 6 sacks Barley, 4 sacks gravels, 1b 200 bran, 1b 3000

---

[34] Katie Whitaker's essay "The Culture of Curiosity" discusses this aspect of collecting in the seventeenth and eighteenth centuries (75–90).

[35] The 13th Earl of Derby's interest in natural history was partially indebted to his father's avid interest in the subject. His father, the 12th Earl, died in 1834 whereupon his son, Lord Stanley, assumed the title. Throughout his lifetime the 13th Earl of Derby (otherwise referred to as Lord Derby) became a leader in the natural history world.

[36] A thoroughly researched collection of essays on Lord Derby's interest in natural history can be found in *A Passion for Natural History: The Life and Legacy of the 13th Earl of Derby*, ed. Clemency Fisher.

[37] The Earl of Derby was also tremendously interested in cross-breeding and the creating of hybrids, partially to create other sources of meat and partially out of scientific interest.

[38] These letters and notebooks are housed in either the Derby Collection at Knowsley Hall or National Museums Liverpool. In this chapter, they are referred to by their catalogue number.

> Hay and a number of Caffir watermelons of which the Ostriches are very fond. I have built a nice house for the Ostriches, I hope therefore you will get them in good condition. I could not send the other Lionesses as there was not room enough for sheep but I have engaged a passage for them in the Bark "Felicity" bound for London. (National Museums Liverpool)

Sometimes the correspondence reveals surprising facts. For instance, in one letter, written July 25, 1842, a dealer informed Lord Derby that a young "White headed Eagle … will probably arrive in Liverpool on Wednesday morning about 10 o'clock, and if any of your Lordship's people are in town during the course of that morning, they will find the bird at the Adelphi Hotel" (920DER [13] Ogilby to LD, National Museums Liverpool). One wonders what the hotel staff thought of it. One also learns that agents occasionally went down to the Liverpool docks to see what various foreign vessels had brought with them for sale. On April 29, 1835, for instance, John Thompson, the superintendent of the Knowsley aviary, informed Lord Derby: "I went to Liverpool in search for the Chinese Vessel thinking to find sumpthing [*sic*] afresh but nothing could I find, there was but one Chinese vessel arrived in Liverpool of late & that had nothing alive" (The Derby Collection, Knowsley Hall).

At times, Lord Derby expressed interest in only acquiring live specimens. In a note to Fry, he declared: "My object is not so much the dead as the living animals" (June 5, 1851, The Derby Collection, Knowsley Hall). But, of course, as discussed earlier, collecting these living and breathing creatures attracted trouble and invited severe disappointment. In spite of his rank, authority, and care, Lord Derby was to suffer many a setback. His notebook pages and letters record innumerable disasters: a storm swept several reindeer overboard (November 23, 1839, letter from Llewelyn Lloyd); an alligator consumed one deer collected in Belize (John Bates's diary September 1843, National Museums Liverpool), and all the spur-winged geese and whistling ducks, collected by Thomas Whitfield at the Cape, expired on the passage back to England (August [(?)] 1843, V.3, Derby Letterbooks, National Museums Liverpool). Sensitive to these catastrophes, Clemency Fisher, curator of Zoology at the World Museum, speaks of the voyage home as "a nightmare,"[39] and goes on to list additional instances of loss and disorder, such as the time the crew of the *Cockatrice* ate the birds destined for Knowsley, and the occasion when "A Vicuña had to be forcibly restrained from eating tobacco and paint." She also recalls the voyage on which "the Llamas refused to eat maize and starved themselves to death, although they did eat the straw used for packing wine" (Fisher 90).

Because of his close connection to the London Zoological Society, Lord Derby also periodically visited the zoo in Regent's Park in order to monitor the fate of its collection. Here too difficulties intruded. Lord Derby's notebooks of "Daily

[39] Steamships did not begin to operate until just before the 13th Earl's death; hence, the sea crossings were slow and arduous.

occurrences" at the zoo are punctuated with announcements of the demise of its creatures, particularly those he had contributed: a stolen gray-breasted parakeet, a copyu rat killed by the watchman's dog, the deaths of a Himalayan deer, a leopard, and a hybrid zebra (National Museums Liverpool).

Lord Derby's risky fascination with the living, however, did not noticeably diminish his lasting attraction to skins, especially those that contributed to his understanding of taxonomy.[40] His agents sent back crates full of them. As part of this quest, Lord Derby not only purchased skins *per se* but also demanded that if a bird or animal died in passage, the collector should skin it. Following such instructions, one of his dealers, John Walpole Wills, wrote: "I have desired in case the Ant-Bear dies, that he [the assistant] may stuff it, in order that *alive or dead* it may reach your Lordship" (November 17, 1834, National Museums Liverpool). Many of his letters regarding the gathering of skins express his wish to gain a more complete taxonomical picture. One written about the skin of a "white lion," sent in 1839 from South Africa, reflects not only this desire but also the collector's yearning to possess the unique object. In the letter, Rev. Fry informs Lord Derby:

> I have bid high for a white lion. I think, in fact, I am sure it is a new species. I hope to send your Lordship a magnificent skin Amral has with him. I have seen it, unfortunately the Missionary who has brought Amral to Cape Town has enduced the poor fellow to promise it to the Governor, he now regrets having done so and as the skin has not yet reached Government House I have still hope. The lion is not *absolutely* white but only so as compared with the Black maned and the yellow maned. It is a large species judging from the skin … If I do not succeed in getting his skin will send a bag, much injured, but made of the skin of a similar lion at least so says Amral. I should be highly delighted to be the means of adding a new species of Lion to our catalogue of Cape mammalia … (National Museums Liverpool. Typescript of letter)

Lord Derby's dealings with agents and dealers, as well as with museums,[41] continued throughout his lifetime. Consequently, at the time of his death in 1851, the Knowsley Hall museum (what Lord Derby referred to in his memoranda for his will as the "museum of dead Subjects") was chock full of skins arranged in cases according to taxonomic groups. It contained 611 stuffed and 607 unstuffed quadrupeds; 11,131 stuffed and 7,700 unstuffed birds.[42] A

---

40 Significantly several species were named after Lord Derby. See "Species Named in Honour of Edward Stanley, 13th Earl of Derby" by Clemency Fisher, Malcolm Largen, and Ian Wallace (Fisher 121–35).

41 Lord Derby acquired many valuable specimens from the sale of the Bullock Museum (Liverpool) holdings, the Leverian Museum (London), and from Dr. Adam Smith's African collection.

42 Clemency Fisher describes the museum as it was during Lord Derby's lifetime: "Most of his specimens were mounted and arranged in conventional glass cases according

staff carried out the necessary care of these specimens, and, significantly, was alert to a skin's vulnerability as well as to the danger of insects described above. Fisher's enumeration of the staff's duties mentions their attempts to preserve the museum's contents:

> He [Lord Derby] was assisted by Wilson, who did jobs like cleaning display cases, writing labels, putting eyes in mounted birds and birds on stands. Louis Fraser became the first curator at Knowsley, transferring (probably in 1845) from the museum of the Zoological Society of London. Fraser's Museum Report from April and May 1849 gives us some idea of the duties of the museum staff at Knowsley: "Dosing Wild Boar with Corrosive Sublimate," collecting camphor in blocks and cutting it for the skin boxes, cleaning "turpentine glasses." (49)

These methods were in part successful, for today in the basement of the World Museum, Liverpool, one can view Lord Derby's cabinets of skins and gaze at groups of carefully labeled and still intact specimens, each carrying with it a record of its provenance and type. These continue to be deemed valuable. During the Second World War, they were among the items temporarily removed from the Museum to protect them from possible bomb damage.

Elizabeth Hornby (1825–1923)

If it were not for a handful of letters scattered among the 13th Earl of Derby's papers in the Liverpool City Library and an illustrated diary housed in the National Maritime Museum, Greenwich, Elizabeth Hornby would have long ago slipped into oblivion. Nowadays, in spite of these remains, she is almost forgotten. Only a few curators, such as Clemency Fisher and Christine E. Jackson, recognize her name and are aware of her experiences at sea and as a colonial resident in South America. She neither rests on the tip of the Victorian specialist's tongue nor figures among lists of traveling Victorian women. Furthermore, no published narrative exists either to prolong her experiences or defy her journal's fading ink. She has never enjoyed the significant attention still given to her uncle, the 13th Earl of Derby. I am among the few, if any, who have ever read through her 400-plus-page diary. The reality is that the ravages of time are working against the possibility of anyone else looking at this carefully rendered account of her life in South America, for so many of her words are vanishing from its pages—gatherings of leaves are now already illegible.

It would be unfortunate if Hornby's writing were to disappear, for her travels in South America proffer an intimate glimpse of a natural history collector's life in the mid-nineteenth century. Significantly they also demonstrate that in Victorian

---

to the taxonomic groups and were set off by linen backdrops. There were also cabinets with drawers of skins" (49). As mentioned at the beginning of the chapter, he left the cabinet skins to the City of Liverpool.

Britain skinning or taxidermy was not an art, as one might presume, exclusively practiced by men and boys, but one occasionally executed by women and girls as well. The seemingly unladylike motions of stuffing the specimen's mouth and wounds with cotton wad so as to absorb the leaking blood and mucus were an integral part of a collector's work. So too were the inserting the knife into the body, making an incision from neck to rump, loosening the skin, pulling it over the head—if a bird, being careful not to damage the feathers, breaking the wings or limbs, and removing the eyes. None of these acts was necessarily alien to female collectors. For instance, late in the century, *The Ladies Manual of Art* (1890) printed a page showing an elegant woman admiring mounted birds, under which are instructions about skinning and stuffing (see Morris 338). Nineteenth-century taxidermy manuals also make it abundantly clear that the practice extended beyond the borders of masculinity. Mrs. R. Lee's 1843 book on the art, already referred to in this chapter, is one piece of evidence and so too is J. Gardner's *Bird, Animal, and Fish, Stuffing and Preserving: A Manual of Taxidermy for Amateurs* (1866), which encouraged everyone to have a go at stuffing birds and, especially, their dead pets.

> I love it [a little bird], prized it dead—'tis with me even now;
> Mute, but still beautiful as when it sang upon the bough
> So, if you ever have a pet—no matter what it be—
> Dog, cat, or bird, or squirrel—just take this advice from me—
> Learn how you may preserve the form which in its life it bore,
> Tis well to keep the memory green of dear ones gone before. (Gardner 7)

Hornby's letters from South America as well as her diary, kept while abroad, suggest she was periodically engaged in this sort of work. She speaks of stuffing two "horrible spiders"—their "legs are full two inches long" (July 12, 1848, 920 DER [13] 1/85/4, National Museums Liverpool)[43]; later, she complains of losing "a capital bird stuffer" who had deserted ship, so she was going to have to teach another sailor how to finish a penguin (920 DER [13] 1/85/7). Apparently she had to complete the task herself, for she wrote: "very busy in the morning finishing the Penguins the little joiner having deserted" (Diary 128).[44] Other diary entries record that on a particular day she was stuffing a sparrow which had recently died (Diary 233); that she stuffed a partridge and "succeeded very well" (Diary 314); later, after the death of a cygnet and pigeon, she proudly noted: "I skinned the former very successfully" (Diary 381–2). Many of these references place the skinning within a domestic context: she often chooses to remain at home, with her

---

43 The letters from Elizabeth Hornby to her uncle, the 13th Earl of Derby, are housed in National Museums Liverpool. When possible, they are identified by their class number.

44 All references to Elizabeth Hornby's journal housed at the Caird Library of the National Maritime Museum, Greenwich (PH/308) are referred to in the text as "Diary" and are followed by the page number.

mother, aunt, and sister, while "drawing, talking, stuffing birds etc." (Diary 349); she gleefully speaks of devouring "much gingerbread" after stuffing some birds which her brother has shot (Diary 381); and, on another day, casually remarks, "I stuffed one of the black & yellow … Birds before breakfast" (381).

Elizabeth Hornby's travels and adventures in skin begin in 1847 when she was just 22 years old. With her family, she set sail from England on board the *H.M.S. Asia*, a ship under the command of her father, the Rear-Admiral Phipps Hornby of the British Navy. During this time, her father commanded the Pacific Station in South America where the British Navy maintained storeships and sheds in Valparaiso, Chile, as well as in Callao, Peru. Because of severe winds and political instability, the Royal Navy thought these locations inappropriate, and subsequently was searching for more secure areas.

On January 6, 1848, the Hornbys embarked from Portsmouth for Chile and Peru, where they were to reside for around three years before returning home on May 16, 1851. While in South America, they periodically sailed to Central America, probably as part of the Navy's search for more appropriate docking areas. While on board the *H.M.S. Asia*, Elizabeth faithfully kept a diary, which she illustrated with her watercolors and which she had bound in leather with her nickname, "Pussy," embossed in gold. This journal records her life as a resident colonial as well as the delights, the tedium, and the disasters of being at sea, including chilling moments when a young naval cadet died: she cut off his hair to send to his mother (Diary 415). Just as poignantly, she mentions catching sight of a Spanish slave ship, a reminder of the suggestive parallel between the trading of skins and slaves. Once on shore, she refers to other harsh realities, such as the cholera epidemics, earthquakes, and the Indian raids in Concepcion.[45] Many of the pages, however, are devoted to the lively social activities associated with a privileged colonial life in Valparaiso: the dances, the dinner parties, the visits to others in her rank, church attendance, cricket matches, horse races, whist parties, band concerts, operas, as well as the required drawing, Spanish, and French lessons.[46] At times she is carefree, and, at others, suitably shocked, especially when she catches "Mr. Lebecan kissing Miss Wulff!!!" (Diary 381).

In spite of her privileged position, Hornby, however, did not confine herself to the polite drawing room or cabins assigned to the Admiralty. Instead, on land, indulging her daring, she repeatedly charged out on horseback and joined many a galloping rough ride, which infrequently ended in a fall. Replicating Darwin's fascination with the gauchos of Chile, she expressed her admiration for them and attempted to learn the techniques of lassoing. In May 1848 a breathless entry

[45] In a November 21, 1849 letter, she describes the Indians: "They are great savages and carry off all the wretched women and Cattle and murder the men" (920 DER [13] 1/85/12).

[46] When Elizabeth Hornby attended one dinner party, she complained that she was sitting next to "a most odious fat yankee" (Diary 356), and after going to the opera, she remarked, "I was glad when it was over" (Diary 390).

exclaims, "we had a great lasso practice" (Diary 101). Indeed, if her diary is to be believed, she was far more successful than Darwin, who, when he tried to learn the skill, managed to entangle and trip the horse he was riding with his lasso. Her pleasure and indomitable curiosity, even her delight in disorder, suffuses these pages.

Figure 2.8 "Ah me! My uncle's spirit is in these stones"
Source: Hornby; National Maritime Museum, London.

Although these moments were thrilling, they never replaced Hornby's consuming interest in natural history. Her letters and diary frequently articulate her enthusiasm for acquiring birds, insects, butterflies, shells, and a few mammals, especially those still zoologically and botanically uncharted. These engaging moments happily disrupted her social calendar. She keenly describes the "pretty birds" the naval officers shot for her (Diary 112), the boating parties which went out in search of black-necked swans and wild ducks (Diary 183), the "long mornings" in which she arranged insects (Diary 204), and the expeditions up a river to bag "splendid Macaws" (Diary 311). She also repeatedly remarks upon visits from others who have a similar interest. For instance, in the spring of 1849, she noted: "Captain Hayes called and brought some beautiful specimens of dried plants, birds, shells and Insects for us" (Diary 212).

Hornby's attention to collecting specimens was, of course, by no means an unusual phenomenon. In addition to being encouraged by the popularity of the pursuit, her interest was indebted to the naval culture of which she and her family were an intimate part. As I have already mentioned at the beginning of this chapter, since the seventeenth century, medical officers and surgeons on board had doubled as naturalists who brought back flora and fauna from new regions so that they could be catalogued and assessed by scientists at home. Moreover, sailors had often amassed specimens, both dead and alive, and had continued to do so for extra income. Dealers were often waiting for them at the docks. This culture supported Hornby's eagerness to collect. Her letters and diary speak of the Assistant Surgeon who got her a partridge (March 25, 1848 in 13th Earl's Letterbooks V.8, National Museums Liverpool), of the surgeon from another ship (the *Cockatrice*) who told her about "a kind of Penguin which comes on the bay" (920 DER [13] 1/85/4), of sending home birds in a special case via the *Sampson* (920 DER [13] 1/85/6), and of sailors shooting birds for her (920 DER [13] 1/85/15). On other occasions, she speaks of a ship's doctor giving her a lecture on anatomy (Diary 14), and of visiting other vessels in which officers and surgeons display their natural history collections. She describes going on shore at Rio to see "a splendid collection of Brazilian birds and insects belonging to Mr. Reeves" (the agent to the English Packet Company in Brazil). She adds, the collection "made my mouth water" (Diary 52). In May 1848 she also records going below to the gunroom, in one of the French boats docked at Valparaiso, to "see some beautiful stuffed birds and shells which some of the officers have collected" (Diary108); on other occasions, she boarded the *Inconstant* to see "some beautiful birds, Pigeons, Parrots ... from New Guinea" (Diary 349), and looked at "some very good insects of the French doctor's" (Diary 356) on board the *Algerie*.

By far, however, the most significant precedent for her attachment to capturing, skinning, and stuffing specimens was her relationship to her uncle, the 13th Earl of Derby. The families were intimately connected: her father's sister, Charlotte Margaret, was married to the 13th Earl of Derby; moreover, her father's mother was the sister of the previous Earl. Because of this close family alliance, the Hornbys spent several months at Knowsley Hall each year; indeed, before sailing for South America, they had been together at the estate. As a young child she would have caught the enthusiasm for natural history and would have learned much about collecting and preserving from Lord Derby and the various people who helped him. She dreamed of having her own collection. On August 29, 1850, she wrote to her uncle to inform him: "I am sending you by the *Inconstant* a few specimens of Birds skins. Any that you have not already got you are welcome to, but if they should be duplicates will you return to Windham to be kept till my return as they make a considerable hole in my collection" (920 DER [13] 1/85/16).

The bond of natural history between Hornby and her uncle is evident in the numerous letters she sent to him from South America. Reading through these, one learns that she devoted much time to procuring specimens not only for herself but also for Lord Derby. On his behalf, she collected and shipped skins as well as live specimens in vessels leaving South America.[47] On November 23, 1848,

[47] Clemency Fisher notes: "Elizabeth herself acquired two of these [black-necked] swans, both females in 1850. One was suffering from cramp, so spent some time recovering

for instance, she wrote to her uncle: "By the time this [the letter] reaches you the ship will probably be just arriving, by her we have sent you a deer skin which was brought us from the Straits of Magellan … I shall be very anxious to hear if the Swan [black-necked swan] gets safe to Knowsley. The Penguin I am afraid will hardly live through the voyage" (920 DER [13] 1/85/5). And in January 1849, she wrote him from the Bay of Coquimbo to announce that she was trying to send him, via the *Calypso*, a pair of red-breasted starlings. She added, "and if I can procure two Coypu rats they shall accompany them" (920 [13] 1/85/8). Throughout, Hornby has got her eye on what her uncle might like to add to his collection. Because of her commitment, she plunges into the hunt; she suffers from an infected bite from a hawk; and she chases after escaped condors: "The condor escaped & we found him walking about the gardens when we came from church. We caught him at last with a noose though he pecked very savagely at us" (Diary 327). She also makes up a recipe of swan soup (oatmeal, bran, and frogs) to nourish the black-necked swans before their journey to England; she dresses the deer in warm flannel clothing to protect him from the cold of the voyage; and she tries to keep birds alive by either nestling them in her "buzzim" (Diary 415) (this attempt failed) or administering pills of raw meat. Ecstatically she exclaims that it was "a good day in the Zoological Department" when she acquires a black-necked

Figure 2.9 "Quiere comprar un Condor"
Source: Hornby; National Maritime Museum, London.

in her mother's bath, but afterwards lived happily in a tank in the garden" (90).

swan and 11 eggs (Diary 370). It is significant that one of her diary's final entries records the fact that after 29 days at sea, upon *H.M.S. Asia*'s arrival in England on May 14, 1851, one of Lord Derby's agents was at the dock waiting to pick up the specimens: "A man from the Zoological Gardens arrived to take charge of the creatures for Knowsley" (Diary 433).

Sections of a letter written to Lord Derby from Valparaiso on June 28, 1850 are worth replicating, for they catch Hornby's intense commitment to these activities; they also reflect her vivacious spirit:

> I think I will claim a Pension for all wounds I have received from various birds & beasts since I came aboard. If some of the family are not killed by the Condor it will be a wonder ... He got loose ... one of the Midshipman ... went in full chase after him pursued by all the men & dogs ... at last after a good chase he was caught in a Lasso by a Chilean ... Captain Johnson of the *Driver* brought me down a deer from California but it unfortunately died on the passage. He has promised however to send me his skin.

In the same missive after going on to inform him that the cold weather had killed the ducks, she continues: "Mr Campbell shot me one curious bird something like a long legged woodcock but I cannot make out what it is. However the skin is in very good presentation as I daresay in time I shall find out" (920 DER [13] 1/85/15).

As the above details reveal, throughout the several years of collecting, Hornby experienced not only its pleasures but also the disorder accompanying it. Ineptness, accidents, upsets, cold weather, casualties, as well as predators were close and threatening companions. Reading through her letters and diary, one learns that a handsome bird was shot, but dreadfully damaged, so lost as a specimen; that a boat carrying two coypu rats upset in the river and "in the scuffle righting her" a member of the crew, "the unfortunate wretch," was killed by a blow on the head and the rats drowned (January 23, 1849 920 DER [13] 1/85/8). The disasters continued. On April 4, 1850, a barge sank and carried with it "three sheep, three Turkies, and nine dozen fowls" (Diary 310), and on August 5, 1850, a distraught Hornby noted that a "poor little Plover," carried in a cage on a horse's back, was swept away in the river's current (Diary 344). More than aware of the dangers in shipping specimens back to England, she mentions the fate of the specimens on board the *H.M.S. Asia* on their return passage: her sister's parrot ate one of the "little Partridges" (Diary 389); a chinchilla jumped overboard and "was drowned or at best died of the cold" (Diary 402); and one of the large cygnets expired (Diary 411). At sea, either heat or extreme cold also caused disasters. An American deer died suddenly in the night (Diary 415); in the extreme heat when docked in Rio, her "last remaining Plover" died (Diary 415), and a few days later, so did two Rio blackbirds.

After becoming aware of the nightmares attending the shipping of live specimens, it is not hard to understand why Hornby frequently relied upon the seemingly manageable unit of skin. But here too, as we have seen, accident, ineptness, and loss compromised control. Gunshot damaged the skins of beautiful birds; skins were lost at sea; and those which survived were often inaccurately mounted. Upon two occasions after visiting museums in both Lima and Santiago where she had gazed critically at the displayed misshapen and rotting specimens, Hornby registered her disgust with these deformities. In Lima, she exclaimed: “The birds are dreadfully stuffed & arranged” (Diary 285); in Santiago she thought the museum “better than the Lima one but the birds are viley stuffed” (Diary 361). She recognized, as did others, that the envelope of skin was vulnerable; it was still subject to disaster. And although it promised to help fill in the gaps or holes within a classification system, it still left room for chaos and disorder. Like the collectors themselves, it rested uneasily “between the poles of disorder and order” (Benjamin 60). For all this, however, both Hornby and, especially, her uncle persisted. Their curiosity (in addition to their occasional sentimental attachment to their collections) combined with opportunity so that they amassed hundreds of skins. For them, natural history was not, as Philip Gosse claimed, “too much a science of dead things” but rather was a pleasurable, exciting encounter with living creatures from places beyond the borders of their immediate surroundings (Barber 41). Indeed, although their amassing of specimens was undoubtedly partially linked to the colonial imperative, it primarily was attached to their genuine admiration for the animals as well as their desire to add to their knowledge of natural history.

Chapter 3
Stuff and Nonsense: Skin and Victorian Animal Portraiture

Introduction

"The Story of the Four Little Children Who Went Round the World"

Fifteen years after Elizabeth Hornby, with her father (Vice Admiral Phipps Hornby), her mother, aunt, and siblings, returned to England following their extensive travels in South America, Edward Lear composed "The Story of the Four Little Children Who Went Round the World" (1867).[1] Although Lear dedicated this piece of nonsensical fun to his close friend "Gussie's" young nieces and nephews,[2] he seems, at least in the writing of the story, to have been thinking more specifically of the Hornbys' adventures on the *H.M.S. Asia* as well as this family's intimate attachment to Knowsley Hall, the Earl of Derby's estate just outside of Liverpool. Indeed, it was at Knowsley that Lear met the Hornbys. Between 1831 and 1837, Lear had either visited or lived at Knowsley Hall in the employment of both the 12th Earl of Derby and the Earl's son (then known as Lord Stanley), who was to become one of the most generous and influential patrons of the natural sciences in Great Britain.[3] (As I explained in the previous chapter, in 1834, upon the death

---

1 For an account of the Hornbys' travels, see previous chapter. To read "The Story of the Four Little Children Who Went Round the World," see Jackson 91–106.

2 "Gussie" is the nickname of Augusta Bethell (1838–1931). In 1862, when she was 24, Lear began to take particular notice of her. (He had known her since childhood.) By 1866 he seriously considered the possibility of marrying "Gussie," but supposedly her sisters disapproved and Lear lost his nerve.

3 Both the 12th Earl of Derby and his son shared an avid interest in natural history. The son especially was to become a leader in the natural history world. As I explained in the previous chapter, he was elected Vice-President of the Linnaean Society in 1817 and was President from 1828 to 1834. He was also a founding member of the Zoological Society in 1826 and its President from 1831 until his death in 1851. It was through his association with the London Zoological Society that he first met Edward Lear and admired his lively and accurate bird portraits. At that time Lear was preparing his volume of parrots (*Illustrations of the Family of the Psittacidae, or Parrots* [1832]), drawing illustrations for Jardine, and working for the bird illustrator, and former taxidermist, John Gould. At Knowsley, Lear became acquainted with people such as the painter Joseph Wolf, John James Audubon and his son, as well as various taxidermists, curators, and global dealers who worked for Lord Stanley. At one point Lear almost accompanied Audubon on his bird collecting travels in

of his father, Lord Stanley acceded to the earldom and became the 13th Earl and was known as Lord Derby. The switch of names can be confusing; for the sake of clarification, in this chapter I shall always refer to the 13th Earl as Lord Derby.)

Lear was at Knowsley because he had been commissioned to paint watercolor portraits of the animals and birds enclosed within the estate's 170-acre menagerie as well as those exhibited in the Knowsley Museum.[4] Initially Lear was considered to be no more than an employee, but after catching the attention of the Earl's grandsons who enjoyed being entertained by his nonsense verses (later to be part of his 1846 *Book of Nonsense*), Lear was asked to dine with the family. Although at first this invitation was confusing to Lear (on one occasion he did not know whether he was a guest at his lordship's table or the housekeeper's), over time, he became friends not only with his employers but also with the new Lord Derby's extended family, among whom were Elizabeth Hornby's parents (her mother was Lord Derby's sister) as well as Elizabeth and her siblings, who annually spent extended summer months on the estate.[5] These friendships endured: in 1841 Lear took a walking holiday with Phipps Hornby; furthermore, the Hornbys as well as Lord Derby helped patronize Lear's later travel writing and landscape painting. Lear, though, as we shall see, never completely forgot his marginalized status.[6]

From its very opening "The Story of the Four Little Children Who Went Round the World" draws upon Lear's immersion in the Knowsley Hall culture, devoted as it was to studying, collecting, exchanging, identifying, breeding, and preserving animals and birds from various parts of the globe. As discussed in the previous chapter, the Knowsley menagerie with its thousands of specimens

---

North America. However, it seems that Audubon became aware that Lear was not robust enough so the invitation was carefully withdrawn. (See Peck, "The Natural History of Edward Lear" 52.)

4 Many of these illustrations appeared in the exquisite and rare *Gleanings from the Menagerie and Aviary at Knowsley Hall*. Robert McCracken Peck gives a sense of the menagerie's size: "The Knowsley menagerie, which Lear came to know intimately, would eventually include several thousand specimens representing 619 different species of birds alone. Among these were 114 species of parrot, 52 species of game birds, 51 species of raptor, and 60 species of wildfowl. The outdoor facilities in which the birds and animals were kept and where Lear spent so much of his time painting from life, eventually covered an area of 170 acres and required a staff of thirty to maintain. The living creatures were complemented by an extraordinarily comprehensive natural history library … [There was] a collection of mounted and preserved birds and mammals which numbered almost 20,000 specimens by the time it was dispersed in 1851" (as quoted in Fisher 43).

5 The family relationships were more complicated than suggested. Phipps Hornby's mother was a daughter of James Smith-Stanley, Lord Strange, and sister to Edward Smith-Stanley, 12th Earl of Derby. Hornby's sister Charlotte Margaret later married her cousin Edward Smith-Stanley, 13th Earl of Derby. This close association between the Earl of Derby and the Hornby family played a significant part in Phipps Hornby's career and politics.

6 For a reproduction of a cartoon showing Lear and Phipps Hornby setting out together from Knowsley Hall, see Clemency Fisher's *A Passion for Natural History*, 94.

was an almost larger-than-life expression of a more general impulse in England to amass and display exotic specimens gathered through trade routes opened up by colonial expansion. In the story, Lear whimsically recasts the adventures of the innumerable curators and collectors who worked for Lord Derby, including Elizabeth and her father during their time in South America. Representing this ethos, the story's children, Violet, Slingsby, Guy, and Lionel, sail across the sea from land to land, island to island; they admire new species, such as the "Co-operative Cauliflower" which hurries off in a "plumdomphious manner" (Jackson 103), explore foreign landscapes, and survive adventures in the wild. Occasional moments in the story obliquely replicate details from Elizabeth's diary and letters. For instance, the narrative humorously evokes her longing to extend the voyage on the *H.M.S. Asia* so as to see more of the world;[7] it also calls attention to the fact that the boat is steered by "Pussy," a detail that recollects recalls Elizabeth's nickname, "Pussy," embossed in gold on the cover of her travel journal. And when the cat on board destroys the exotic parrots by biting off their tails, what jumps to mind are the many entries, written during the *H.M.S. Asia*'s return voyage, in which Elizabeth complains of her sister's cat preying upon and sometimes devouring the live specimens she and her father were attempting to take back to the Knowsley Hall Menagerie.

In the mode of nonsense, which typically undoes the threat of danger, Lear delightfully stretches the truths he is representing throughout this story so that toward the end of the narrative when the children lose their sailing vessel to the jaws of a ferocious aquatic creature, they improbably, for 18 months, travel home across land on the back of an elderly rhinoceros, which is also transporting "a crowd of kangaroos and Gigantic Cranes" (Knowsley was famous for its kangaroos and cranes; furthermore, captured animals were often transported on the backs of large animals). Once they all arrive in England, however, this obliging rhinoceros does not fare so well: he is summarily slaughtered, flayed, and stuffed. The narrative brusquely concludes: "As for the Rhinoceros, in token of their grateful adherence, they have him killed and stuffed directly, and then set him up outside the door of their father's house as a Diaphanous Doorscraper" (Jackson 106)—only in the topsy-turvy world of nonsense, it seems, can a rhinoceros's celebrated thick skin become "diaphanous" or transparent.

---

7 In a November 21, 1844 letter to her uncle, Lord Stanley, Elizabeth writes: "I wish they would order us home by India for now having gone abroad I think it would be pleasanter to go right round the world than only half round and back again" (920 DER [13] 1/85/12, National Museums Liverpool).

Figures 3.1/3.2 From Edward Lear's "The Story of Four Little Children Who Went Round the World"

Lear's harsh, abrupt conclusion immediately empties "The Story of Four Little Children Who Went Round the World" of its harmless, humorous play, and baldly exposes the fate of numerous grand, exotic mammals whose bodies were mounted and exhibited by hunters and scientific institutions. The exaggerations of nonsense disappear. Lear's rhinoceros is yet another victim of the self-appointed license to show off colonial authority.[8] It shares a place with those trophies prominently

[8] It is interesting to note that occasionally this imperative to demonstrate dominion was extended to those with fewer means, who were encouraged by taxidermy manuals to capture and stuff something small or ordinary, like a squirrel, in order to demonstrate their bravery and command. A poem in J. Gardner's 1866 *Bird, Animal, and Fish: Stuffing and Preserving*, a manual composed for amateurs, encourages such behavior. In the poem, the stuffed squirrel speaks:

It took six boys to catch me;
 And then I bit them so,
That they were forced to choke me,
 Or else to let me go.
And now I'm dead they've stuffed me,
 To let all people know
How brave and active they were all,

featured at the 1886 Colonial and Indian Exhibition, and joins the company of the 16 tigers shot in India by the Prince of Wales (Edward VII), later exhibited to the public in the lecture room of the London Zoological Society (1877). It also becomes still another trophy rhinoceros head on the wall of that Society's meeting rooms at Hanover Square.[9] More particularly, Lear's rhinoceros is subjected to the imperative to "civilize" or tame the exotic other by converting it into a domestic commodity, in this instance, into an item upon which the paternal colonial might wipe his muddy boots.

Wardian Furniture

The conclusion of Lear's story cannot be dismissed as mere nonsense; in its hyperbolic way, it tells the truth, for the practice of removing an animal part (such as a foot) or its skin and converting that section into something ornamental was almost routine. Shaped into a doorscraper, a utilitarian, subservient object, Lear's rhinoceros is transformed into what was popularly known as Wardian furniture. These furnishings were named after the well-known London taxidermist Rowland Ward, who made a profit by turning animal parts into ornamental household articles. One of the more popular items in his stock was the elephant's foot liqueur stand (the interior of the enormous foot was fitted as a spirit-cellar while the lid contained boxes of cigars).[10] Ward also constructed pieces of furniture out of rhinoceros hide. In an advertisement at the back of *The Sportsman's Handbook to Practical Collecting*, Ward boasts of having, after "six years" labor, perfected the process through which he was able to make the thick rhinoceros's skin pliable for ornamental purposes. He announces that, as a result, the rhinoceros hide can now be worked "not only into sticks, trays and smaller articles of use or ornament but into the construction of Cabinets, Tables, etc. etc." Proudly, the advertisement continues: "There is, for example, among the articles [out of rhinoceros skin] … prepared for His Highness the Maharajah, a beautiful little table of the most exquisite polish and beautiful grain, which looks as if it had been

---

And what six boys could do! (10–11)

As a sideline to this observation, I recently opened a 2012 mail-order catalogue from Wireless and discovered, listed as a "Customer Favorite," a "Perfectly cute and kitschy" Mounted Squirrel Head trophy, "sculptured in super-realistic resin (no squirrels were harmed in the making of this trophy)." On the following day yet another holiday catalogue (What on Earth: A Collection of Funwear & Delightful Diversions) arrived and what should I find but a fake "life-like rhinoceros trophy head" for $39.95: "If you're asked where you got it, say you bagged it yourself. Imagine how exotic you'll seem" (12). Are we "faux Victorians"? In spite of our contemporary criticism and disapproval of taxidermy, we are apparently still drawn toward the aura of power hovering around the mounted trophy.

[9] See [May 23] 1885 satirical *Punch* cartoon, "The Meeting of the Zoological Society, Hanover Square," showing a rhinoceros's head, sporting a frilly bonnet and glasses, mounted on the wall of the Zoological Society rooms (*Punch* 88: 251).

[10] A rarer item was a hall-porter's chair constructed out of a small elephant.

polished out of some beautifully veined *transparent* stone or clouded amber" (179 [italics mine])—perhaps this transparency is the source of Lear's choice of the word "diaphanous" to describe the doorstop's appearance. Yet another established taxidermist in London, Gerrards, also featured ornamental furniture made by "Mounting Hippopotamus & Rhinoceros Feet in Silver, Electric Plate, Brass, Copper & Copper Bronze" (Morris 116). The resulting umbrella stands, ink wells, lamps, and ashtrays fashioned out of the rhinoceros's feet are, to say the least, a strange, distorted form of portraiture.

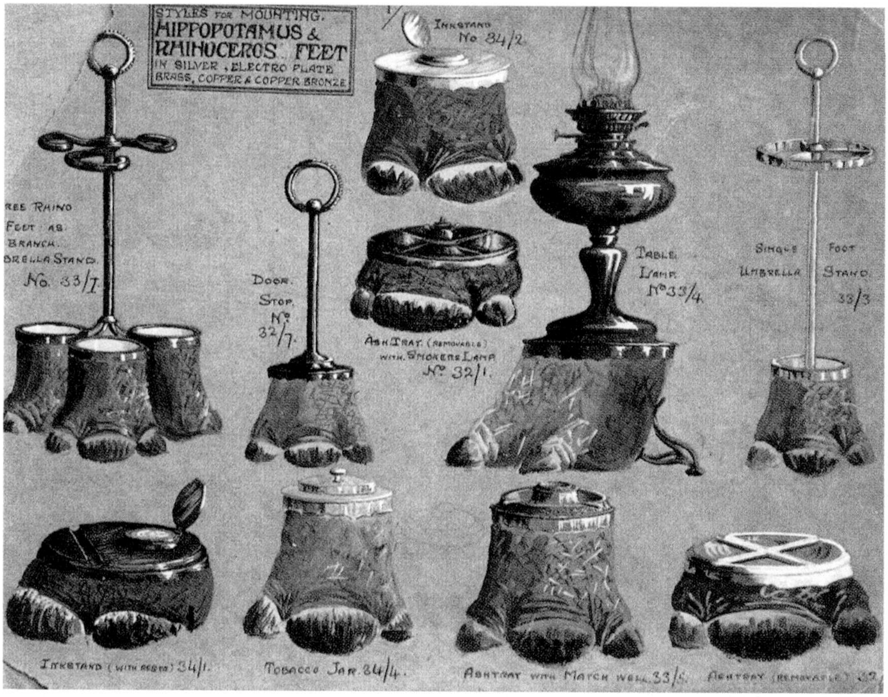

Figure 3.3 Wardian Furniture

However fascinating the transformation of the rhinoceros into a piece of Wardian furniture might be, this rather bizarre Victorian phenomenon is not nearly as absorbing or as informative as Lear's deliberately matter-of-fact reference to the flaying and stuffing of the rhinoceros's skin—especially after such a point has been made of the creature's generosity. The startling ending to "The Story of Four Little Children Who Went Round the World" makes one wonder about the popularly accepted practice of taxidermy and prompts one, more generally and inclusively, not only to think further about the centrality of skin in animal portraiture but also to examine Lear's own position with regard to the importance of skin in his work as a natural history illustrator.

Questions come to mind: what were the Victorians saying about the nature of skin when they chose to fill and shape it into some semblance of what once was living? Why, for instance, in 1829, was John Gould asked to stuff George IV's "pet" giraffe, which before arriving in England had initially traveled from Sonnaar to Cairo on the back of a camel, and why was Joseph Wolf hired to stuff Queen Victoria's pets? Furthermore, why did amateurs and more modest individuals consult taxidermy manuals so that they could preserve a favorite canary? (One cannot help but recall the person who comes to pick up the stuffed canary from Venus's shop in *Our Mutual Friend.*) On a more professional level, why, for example, did the 13th Earl of Derby, as well as many of the zoos, hire taxidermists and illustrators so that they might stock their museums and libraries with mounted specimens and stacks of natural history drawings? Knowsley's museum of "dead subjects" (animals and birds which had died in the menagerie) consisted, for instance, of 611 stuffed quadrupeds and 11,131 stuffed birds. And, what prompted the sailors and naval officers, as well as Elizabeth Hornby, on board the *H.M.S. Asia*, to stuff many of the creatures they had gathered on their travels? Could they have not been content merely to save the skin alone?

Essentially, one wonders, what does taxidermy contribute to an understanding of the significance of skin for the Victorians? Moreover, and perhaps more significantly, what exactly does the common practice of using a stuffed specimen as a model in order to paint or draw a particular mammal or bird reveal about the crucial part skin played in animal portraiture?

In order to address these questions, I shall devote the first part of this chapter to expanding my discussion of the Victorians' fascination with animal skins, and then I shall move on to consider the role of skin not only in the art of taxidermy[11] but also in the rendering of natural history illustrations. As an example of this preoccupation, in Part Two, I shall return to my initial focus on Edward Lear, and discuss his rendering of skin in his animal illustrations. In this section, however, rather than focusing on "The Story of the Four Little Children Who Went Round the World," I shall concentrate upon Lear's immaculately executed watercolor portraits of birds and mammals, done for patrons such as Lord Derby. Lear's paintings and prints of these creatures reflect not only the Victorians' attraction to and dependency on skin, but also the consequential demand that natural history artists pay minute attention to a creature's exterior markings and texture. In Part Three, I shall conclude by exploring Lear's critical reaction, through his nonsense drawings, to this strict requirement. For Lear, it seems, the key to a true portrait of a creature lay not so much in the overpowering details of a meticulously executed skin as in the intimacy of that skin, which allows the portrait to breathe and move. Unfettered by the scientist's classifying gaze or the collector's/colonial's prerogative that transforms an animal or bird into yet another commodity, such as the "diaphanous doorstop," Lear was interested in another kind of portraiture.

[11] *Taxidermy* is based upon two Greek words: *Taxis*: to arrange; and *Dermis*: skin.

Part One: Skin, Taxidermy, and Illustration

Victorians and Skin

As I have established in earlier chapters, even when imperfect or fragmented, skin was considered by the Victorians to be a primary source of identity, the essential signature of difference. Through its markings and colorings, skin offered an immediate visual key to cultural and racial identity, and, as such, instantly exposed both animals and humans to the judgment of others. Animals were particularly vulnerable to this kind of scrutiny, particularly when exhibited outside their natural contexts (in zoos, museums, and estates) and displayed in cages or cases without a narrative or a prop, except perhaps a label, to define their position or reveal their character. These creatures were consequently viewed almost exclusively through the medium of the textures, colors, and patterns of their skins. In a sense, these features provided the defining vocabulary for the Victorians' lexicon of taxonomies.[12] It was therefore not unusual to find William Swainson, in his 1838 *Animals in Menageries*, distinguishing "the tiger of Bengal … from all other ferocious animals," by means of "its beautiful skin, marked throughout with narrow dark stripes upon a yellowish buff ground" (104). Even Richard Owen, who was more interested in the anatomy of natural history specimens (getting beneath the skin), chose not to ignore the skin's surface in order to distinguish one species from another. In his *Memoir on the Gorilla*, for instance, Owen discriminates between a gorilla and a chimpanzee according to the ways in which the respective hairs on the gorilla's and the chimpanzee's epidermis reflected the "bright sunlight."[13]

Skin was not mute. It "spoke" to the Victorians and insisted upon being heard by those who gazed upon it. The brilliant narrow and contrasting stripes of the tiger's fur or the light on a gorilla's hairs created a vocabulary and a voice to which scientists, collectors, and even the general public listened. These markings reverberated in people's consciousness and were eminently louder than any growl, cry, song, or howl; though I hasten to add that occasionally sounds were added to stuffed creatures. One instance I have run across is from a nineteenth-century *Illustrated Guide to the Jungle*, which vividly describes Rowland Ward's taxidermy constructions at the Colonial and Indian Exhibition of 1886. I was fascinated to read that in his depiction of a group of tigers attacking a person, Ward

[12] It would be remiss to overlook the fact that in addition to being interested in skins, scientists were also fascinated by the skeletons and viscera of these animals. For instance, A.D. Bartlett was not only intent on keeping the skin of an "ourang-utan" but also committed to preserving its skeleton and the viscera "to secure accurate information to the naturalist" (*Wild Animals in Captivity* 5).

[13] "[T]he degree of admixture of different-coloured hairs," reveals that a living gorilla, "seen in bright sunlight, would in some positions reflect from its surface a colour much more different from that of a chimpanzee" (Owen 11).

inserted an "apparatus" that sent a "nerve tingling" cry of "'*Bagh! Bagh!*'" and a "despairing shriek" from the tigers' victim (*Illustrated Guide to the Jungle* 33). The assemblage recalls the remarkable painted wooden figure of "Tippo's Tiger" (1793) in which a mechanical organ was placed inside the modeled tiger's body (one turned a handle to activate the organ) so as to capture the despairing sounds of the victim as well as the aggressive growls of the tiger.[14]

The powerful vocabulary of an animal's skin even occasionally helped portray a person's identity. For instance, in *Our Mutual Friend*, Charles Dickens summons the dark, heavy density of a rhinoceros's skin to describe Mr. Noddy Boffin (the "Golden Dustman") so as to distinguish him from the novel's menagerie of characters (the bird of prey, the crocodile, etc.). To illustrate how "thick" and "dense" the illiterate Boffin initially appears to be, Dickens compares Boffin's garments (his second skin: his clothes) as well as his complexion to a rhinoceros's hide. Dressed in a pea overcoat, and carrying a large stick, the "broad, round-shouldered" Boffin wears "thick shoes, and thick leather gaiters, and thick gloves like a hedger's." He displays "an overlapping rhinoceros build, with folds in his cheeks, and his forehead, and his eyelids, and his lips, and his ears." The narrator concludes: "A very odd-looking fellow altogether" (46).

For many Victorians, skin, though, was not simply a language of branding. It also stood metonymically for a being's entirety. In this mode, skin functioned as an index to character as well as a site of identity. Consequently, for William Swainson, the stripes on the Bengal tiger's fur did not merely distinguish the creature from other animals but also revealed the tiger's "savageness and butchery" (*Animals in Menageries* 104). Similarly, the rhinoceros's thick, irregularly folded skin, in addition to being the identifying attribute of this unusual mammal, was sometimes conceived to be the site of the rhinoceros's temperament; it registered his bad temper. In Rudyard Kipling's fable "How the Rhinoceros Got His Skin," the awkwardly layered armor-plated hide records the rhinoceros's failed efforts (his continuous rubbing) to rid his body of the irritating stale cake crumbs lodged under his skin as a punishment for stealing a cake. Continuously exasperated, the rhinoceros develops a disagreeable disposition. Kipling's narrative might be fictitious, but the concentration on the animal's skin as a metonymic text is not.

Human skin, of course, was not exempted from this kind of focus. It too was regarded as a guide to portraying character. A skin's coloring was thought to disclose an individual's disposition, an orientation, which, as several critics have recently noted, influenced Victorian novelists' depiction of temperament. Consider the blush on a cheek or a sudden paleness of complexion, which visibly signifies to the reader the presence of a conscience, and, thereby, reveals, for instance, the moral identity of characters in such novels as *Oliver Twist* or *Mary Barton*.[15]

---

14 "Tippo's Tiger" can be seen at the Victoria and Albert Museum, London.

15 For further and more nuanced thought on this matter of "the blush" see Mary Ann O'Farrell's *Telling Complexions* which explores the frequent use of the blush in Victorian

The Victorians, however, also recognized that skin registers more than the identity or character of what it envelops. It also carries with it, dead or alive, an intimacy, a smell, a feel, which can bring the observer closer to what was or is a living, breathing subject. In South Africa, when the celebrated Victorian hunter Roualeyn George Gordon Cumming lifted and smelt the skin of a dead wildebeest, the skin's "delicious odor of the grass and wild herbs" where the animal had lain evoked the living creature (*Five Years of a Hunter's Life*, 1: 81). And again when Cumming bent over the skin of the eland he had just shot, the skin's "most delicious perfume of trees and grass" recalled the breathing animal (*Five Years of a Hunter's Life*, 1: 253). On more familiar ground, the intimate sensation of touching and smelling the skin of a pet ushered its being into consciousness and affirmed its presence.

This attribute reminds one that in the human imagination, skin is often closer to the texture of life than is any well-constructed skeleton or carefully preserved viscera—recall the piece of elephant trunk attached to Maharaja's skeleton in the Belle Vue Museum (see Chapter 1). For instance, Joseph Wolf, an official illustrator for the London Zoological Society, vividly experienced the inclusive and resurrecting power of skin when he found himself attempting to draw a new species of antelope for the Society from nothing but a "peculiar-looking long, narrow skull, with a label attached to it." Only when Wolf ambled into another room and found the antelope's crumpled skin draped on a "kind of wooden horse" was he able to begin recreating "a life-like image" of the creature (Palmer 240). For Wolf, as well as for many others, this draped hide carried with it the texture of the antelope's life. Though removed from the antelope's body, it was somehow still integrally attached to the being it had once covered.

A few years ago I was reminded of this reality when I was walking around the "Skin" exhibit at the Wellcome Institute, London, and paused before the removed tattooed skins of nineteenth-century sailors. The fact that these pieces had once been attached to living people took me aback.[16] The life, though I had not witnessed it, was uncannily present. The experience was related to what Rachel Poliquin, a recent commentator on taxidermy, calls a raw experience. I should, I suppose, make allowances for the fact that I was gazing at human rather than at animal skins. Undoubtedly there is more self-identification and more at stake. One shudders when one learns, for instance, of the book of sonnets written by a nineteenth-century Russian poet to his mistress, bound in the tanned skin

---

novels to indicate character and inner emotions or desires, and also look at William A. Cohen's *Embodied: Victorian Literature and the Senses.*

[16] Poliquin in her recent study of taxidermy, *The Breathless Zoo*, recognizes the fact that when we are in the concrete presence of what had once lived, we are "haunted" by "the presence of death" that casts "an uneasy shadow" (41). Another interesting study of taxidermy is Merle M. Patchett's 2010 PhD dissertation, "Putting Animals on Display: Geographies of Taxidermy Practice." Patchett received her degree from the Department of Geographical and Earth Sciences at the University of Glasgow.

from his amputated leg, or of a young French countess who, in 1802, wished that after her death, a large piece of skin cut from her shoulders should be sent to her lover and serve as a binding for one of his books (Connor 45). Animal and bird skins are, however, not exempt from this immediacy. They too carry the intimate remembrance of a living being.

What is intriguing to me about the Victorians is their desire to cling to the aura of life residing within a skin after death. Contrary to what Poliquin proposes in her fine discussion of taxidermy, that Victorians, when stuffing a creature, were primarily recognizing the attendance of death, I suggest—and this is perhaps a minor difference—that, instead, they were acknowledging the remnant of life they believed remained within and exuded from the sight and feel of that skin; in a sense, they were reviving what had been lost. Preserved, resurrected, repaired, shaped, and filled to form a mounted specimen in drawing-rooms, interiors of estates, or in crowded museums, the taxidermy specimen realized the vitality within the remaining fur and feathers. As those Victorians who arranged either to mount exotic animals or to stuff their pet canaries (or in Dante Gabriel Rossetti's case, his pet wombat) would attest, these remains extended the life and, depending on circumstances, allowed an individual either thoroughly to study a newly-discovered specimen or privately to retain an attachment to a familiar, beloved pet.[17] Lines from a clumsily written poem in Gardner's 1866 amateur taxidermy guide address the latter possibility. Readers are encouraged to stuff their companion dogs, cats, and birds so they can prolong the breath of skin and keep remembrance alive:

> So, if you ever have a pet—no matter what it be—
> Dog, cat, or bird, or squirrel—just take the advice from me—
> Learn how you may preserve the form which in its life it bore
> 'Tis well to keep the memory green of dear ones gone before. (7)

For the Victorians, skin was more than merely life's envelope; it figuratively conveyed life itself. It was not "breathless" but revitalizing; it exhaled life.[18]

---

[17] On a recent tour of the working spaces behind the public areas of Buffalo's Museum of Science, I was fascinated to discover shelf after shelf of unclaimed mounted pet dogs and cats. The curator explained that these had been there for years. Apparently, pet owners at first thought it a good idea to bring their beloved animal back to life, so to speak, by having it stuffed, but upon reflection, had second thoughts, so had just left them there. When looking at nineteenth-century taxidermy papers, I have been struck by the fact that taxidermists insisted that they be paid ahead of time if someone brought in a pet to be stuffed. The reason for this was evident in my visit to the back rooms of Buffalo's Museum of Science.

[18] My point contradicts Poliquin's sense that taxidermy is "breathless." Appropriately, she entitles her study of taxidermy *The Breathless Zoo: Taxidermy and the Culture of Longing.*

In this respect, I suggest that the art of taxidermy is quite different from Frankenstein's method of creating a being from the inside outward by first collecting parts, assembling them, and then, finally, covering the assemblage, as if tying up the bits, with yellowish skin (perhaps collected in a slaughter yard). Frankenstein begins with the body's interior and works out toward the skin surface.[19] Unlike the taxidermist, he does not commence with the life-giving, life-containing skin. Frankenstein does not acknowledge that existence comes through skin.

Because life, so to speak, begins with and resides within the skin proper, Victorian taxidermists went to great lengths to make certain that they did not abuse it when mounting their specimens and forming sculptured portraits of their subjects. Even though these artisans or technicians, when removing the skin from an animal or a bird, had, almost literally, to invoke a second death by penetrating or violating the skin's boundaries, they made sure that the cut was clean, and not "fatal" so that the fur or the feathers were neither damaged nor discolored by bodily fluids; retaining the skin's natural life-giving appearance was primary. An opening section in Captain Thomas Brown's taxidermy manual addresses this priority:

> When a quadruped is killed, and its skin intended for stuffing, the preparatory steps are to lay the animal on its back, and plug up its nostrils, mouth, and any wounds it may have received, with cotton or tow, to prevent the blood from disfiguring the skin. A longitudinal incision is then made in the lower part of the belly, in front of the pubis, and extended from thence to the stomach, or higher if necessary, keeping in as straight a line as possible, and taking care not to penetrate so deep as to cut into the abdominal muscles. In some instances, the incision is made as high as the collar bone. In this operation the hairs must be carefully separated to the right and left, and none of them cut, if possible. The skin is also turned back to the right and left, putting pads of cotton or tow between it and the muscles, as the skinning is proceeded with. (7)

For similar reasons, taxidermists were also interested in dealing with "good skins," those which had not be ruined by a bullet, a knife, or by a hunter's "coarse," heavy hand, which, as Brown complained, had "disordered," stretched, deranged, and sullied the specimen's feathers or skin and compromised what once had been alive (4). In her letters home to her uncle, Lord Derby, Elizabeth Hornby grumbles about "two handsome" fly catchers that were "so much shot that they are not very good specimens" and could not be stuffed (National Museums Liverpool, 13th Earl's Letterbooks V.8, March 25, 1848).[20] In another missive from Valparaiso she

19 The same can almost be said of Mr. Venus in *Our Mutual Friend*, who assembles bodies out of bits and pieces and articulates skeletons as well. For him skin is not necessarily an inclusive beginning; it is only incidental to what he is constructing. In this respect, Venus's shop resembles Frankenstein's laboratory more than it replicates a taxidermist's.

20 All of Elizabeth Hornby's letters come from National Museums Liverpool. In subsequent quotations, the date or reference number only will be given.

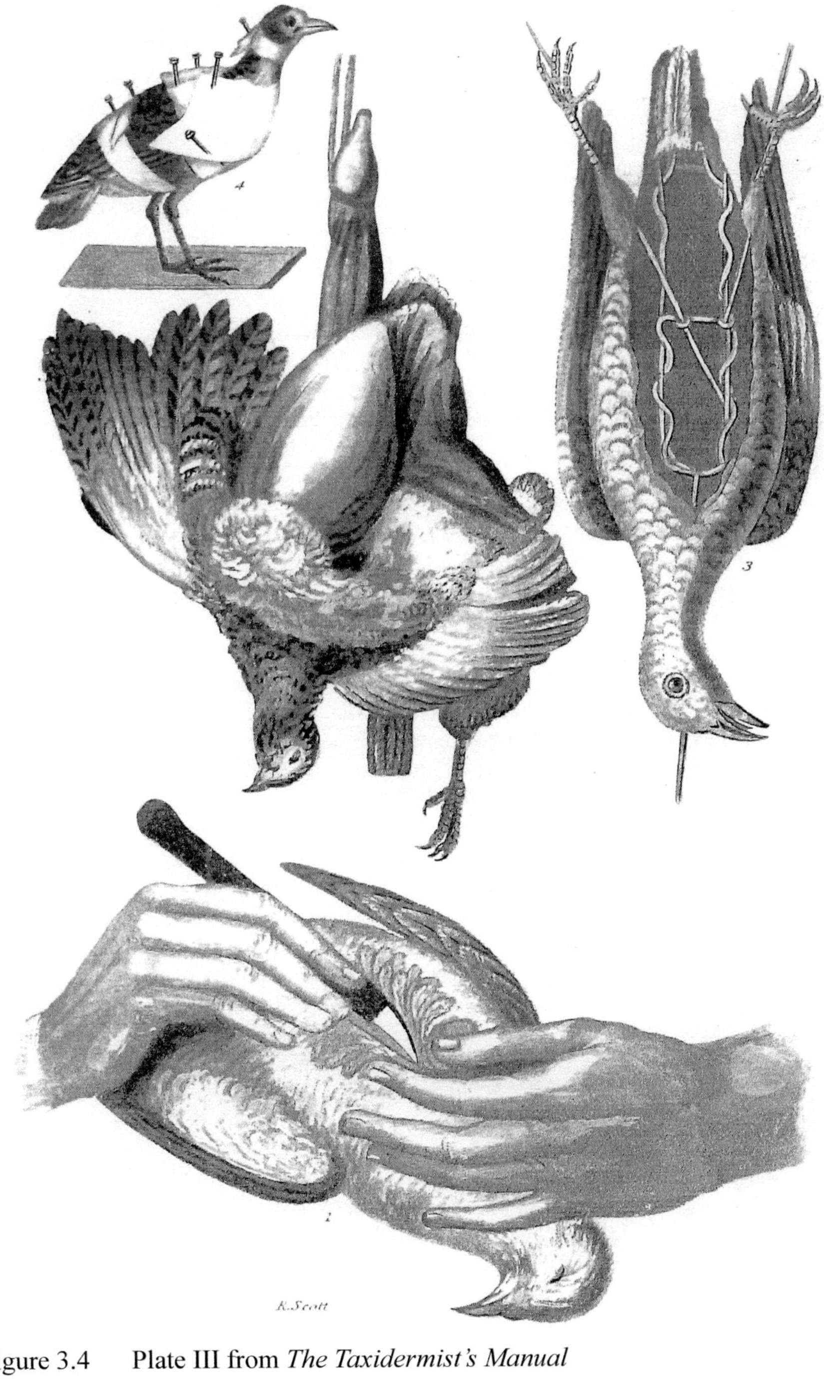

Figure 3.4 Plate III from *The Taxidermist's Manual*
Source: Brown 1833.

also warns her uncle that among the specimens sent back to England are: "Several [birds] … shot by amateur sportsmen" which are consequently "a good deal damaged & … very difficult to stuff" (August 29, 1850, 920 DER [13] 1/85/16).[21]

Once the skin had been removed, the stuffer, as a taxidermist was then called, had to exhibit skill in rendering the animation residing within the skin. There were, of course, many methods to flesh out the life of the skin. One conventional way was to construct the general shape of the body out of wire and then wrap cotton tow around this support in order to replicate the folds, bumps, and swellings of the creature's appearance. The skin was then fitted over this frame.[22] As technology

---

[21] As were other stuffers, A.D. Bartlett, the taxidermist and superintendent of the London Zoological Gardens, was sensitive to this predicament identified by Hornby. He wrote: "The first object of a taxidermist is to render all the damage or wounded parts of a skin as perfect as possible, and this can be done by a skilful operator in such a manner as to render the detection of the damaged parts next to impossible" (*Bartlett's Life Among Wild Beasts* 6).

[22] J. Gardner in *Bird, Animal, and Fish: Stuffing and Preserving* instructs:

> Now comes the "stuffing" and "setting up" process, in which great taste and skill may be displayed. In the first place you must procure six lengths of stout wire (about as thick as twine will do) of ten inches each. Soften the ends of the wires by making them red-hot, and then sharpen the points with a file. Next, make a body of tow, wrapped tightly round the length of one wire. This body of tow you must make as nearly the size of the animal as may be, shaping it into the requisite form, larger in the centre and smaller towards the two ends. Place this false body inside the skin up to the neck, so that the sharpened point of the wire comes through the nose. Pass a wire through each of the fore-feet into the body, and clench each one on the other side. Serve both fore and hind-legs in the same manner, and then proceed likewise with the tail. We have now, as it were, a new skeleton for our stuffed pet. The next process is to fill up all cavities left by the artificial body with loose tow or wadding. Then neatly sew up the skin, so that the stitches do not show through the fur. The proper sort of artificial eyes can be purchased at any respectable bird-stuffer's. They are fastened in their places with wire, or with putty. Eyes of various kinds, especially for the larger kinds of animals have a bit of wire at back. If not convenient to stuff the squirrel directly, the skin, after it has been prepared, can be placed on one side and kept for any length of time. All that is necessary when you want to stuff your specimen, is to damp and soften the skin. (12)

Another description of this process is in William Swainson's 1840 taxidermy manual:

> Commencing with Quadrupeds, the operator should begin by opening, cleaning, and filling the mouth with cotton or tow, to prevent any blood or moisture from exuding. All wounds should be treated in the same manner. The animal is then stretched on its back, and the hairs being tuned to the right and left, the skin is to be opened in a straight line down the middle of the abdomen, commencing from the arch or hollow of the pubis, and ending with the stomach: the upper part of the slit may be extended to the collar bone; but as the operator gradually acquires dexterity, he will be able to decrease or shorten it. Care must be taken not to injure the muscles of the belly, by making the first incision too deep, otherwise

progressed, however, this method was gradually replaced by meticulously measured artificial bodies or "manikins" (made of such materials as wood, papier-mâché, plaster, and even peat) to which taxidermists could bind cotton tow before enveloping them with the treated skin. A plate from John Rowley's *The Art of Taxidermy* illustrates one version of this technique. The photograph shows a manikin for a zebra. In order to see the life-giving effect skin renders, it is helpful to look at the plate which follows, showing the manikin finally covered with the removed and treated hide. For all purposes the stuffed zebra has now recovered a semblance of its life; its skin has made up for what it lost in dying.

This reality was made even more convincing to me when, recently walking around the Zoology Museum at Cambridge University, my attention was drawn to a child gazing at a free-standing stuffed zebra—no cage or glass case enclosed the specimen. I watched the child repeatedly and tentatively extending his arm and fingers (and then hurriedly withdrawing them) as if daring himself to touch the zebra. I identified with the child, for I too was caught wondering whether or not the zebra was really dead. I also feared the specimen might react to my touch. The vibrant and shaped skin caught us both within the paradox of what one might call a living death.[23] We had, as Poliquin so rightly acknowledges, experienced a "compelling strangeness" which "arises in large part from the contradictions between the *perception* of this object on display—mute and manufactured—and the *recognition* that this is no mute and manufactured object" (50). Through the zebra's mounted skin, we had stepped into the "animal zone"—the face-to-face physical presence, the illusion of immediacy one experiences standing close to an animal one ordinarily would not get near without impunity. For me, and I believe for the child, however, this vibrant stuffed skin did not completely remove us, as Poliquin suggests, from the idea of harm (we were not exempted from our vulnerability before a wild beast) nor were we excused from the condition of

---

> the intestines will fall out and soil the fur. The operator then proceeds to separate the skin from the flesh, both to the right and left of the belly, placing pads of tow or linen between, and sprinkling powdered chalk on the flesh as the surface is absorbed: the anus is next detached from the rectum, the tail cut off interiorly at the last joint, and each thigh separated at its junction with the bones of the pelvis. Hitherto the animal has remained upon its back; but it must now be laid on its side, the posterior part towards the right, and give more facility for skinning the back: this last part is always the easiest. For quadrupeds of a small or middling size, it is sufficient to take the skin in one hand, and the body in the other, and by drawing them in contrary directions, to unskin the body as far as the scapulae, or rather to the shoulders. (*Taxidermy, a Bibliography and Biography* 31)

23 Poliquin quite rightly observes: "taxidermy embraces viewers within an aura of wonder. That atmosphere of compelling strangeness arises in large part from the contradiction between the *perception* of the object on display—mute and manufactured—and the *recognition* that this is no mute and manufactured object. Viewers can never escape the startling realization that this static thing in a very real sense is an animal still: the eyes may be glass, but the animal stares back" (50).

Figure 3.5 "Manikin for Zebra, Completed, Ready for the Skin"
Source: Rowley 1898.

time.[24] Although most historians and commentators on taxidermy remark on the fact that taxidermy removes a specimen from the vagaries of time so it can, more or less, be adored, studied, or preserved from extinction, that effect is not always possible. The reality of decay (the instability of skin) compromises that ideal. As a consequence, both the child and I (even perhaps the zebra) remained close to the boundaries of temporality (in the next moment, the zebra might flinch) and fear (in a minute, we might be bitten). As my experience had instructed, when I gazed at the removed tattooed human skin at the Wellcome Institute, the usually and seemingly rigid distinctions between life and death (within what Connor calls this "corpse of a corpse"[25]) became muddled.

My experience in the Zoology Museum reminded me of the 1856 *Punch* cartoon, already mentioned in the previous chapter, featuring a person admiring a private collection of stuffed specimens (see Figure 2.6). The drawing shows "Old Mr J—N—S" regarding what he thought was a stuffed cat, but, much to his "discomfiture" and surprise, the taxidermy rendition of the cat turns out to be a very much alive Great Horned Owl which, upset at being approached (the man had probably tried to stroke it), angrily attacks him. In our own time, contemporary

24 Poliquin suggests, "Taxidermy exists because of life's inevitable trudge toward dissolution. Taxidermy wants to stop time" (7).

25 See Connor 11.

Figure 3.6 "The Specimen Completed"
Source: Rowley 1898.

artists working with the medium of taxidermy often play with this baffling, uncanny phenomenon, when the stuffed skin keeps one unbalanced between life and death. For example, a taxidermy piece entitled "I'm Dead 2007" by David Shrigley shows a stuffed kitten standing on its hind legs and holding a wooden sign or picket, which reads "I'm Dead." This ironic (perhaps, postmodern) announcement, of course, helps the observer avoid "Old Mr J—N—S's" dilemma and mocks the observer's attempts to navigate the visible confusion between life and death.

Skin and Accuracy in Taxidermy and Natural History Portraits

Because the Victorians invested so much in the force of skin, and by extension, the art of taxidermy, one cannot read far in the literature without being interrupted by people's disapproval of badly displayed or treated skins as well as their displeasure of hideously, distorted, and inaccurately mounted specimens. This criticism was not just reserved for the professionals, but also was part of the popular perception. An April 2, 1881 *Punch* cartoon "You have stuffed my parrot very badly" shows a disheartened and dissatisfied customer picking up her stuffed pet parrot from a taxidermist. Badly preserved, the specimen is already losing its feathers. This specimen is an example of what the taxidermist and naturalist Charles Waterton critically referred to as "death in ragged plumage" (325).[26]

Some of the art's most vociferous critics were the taxidermists themselves who complained of "ragged moth-eaten old specimens … stuffed by some incompetent hand" (Browne, *Artistic & Scientific Taxidermy* 12), "hideous" monkeys (Buckland 319),[27] or misshapen creatures, and even forgeries or falsehoods such as the time Du Chaillu was accused of painting a gorilla's face to cover up a wretched skin.[28] Some even spoke of taxidermists as being nothing but "upholsterers." Elizabeth Hornby was part of this culture. When visiting a museum in Santiago, she complained in a letter to her uncle: "The creature [Huernal?] is very badly stuffed indeed … The other specimens of natural history were miserable. Viley stuffed & badly arranged" (October 25, 1850, 920 DER [13] 1/85/17). These accusations were often aired in public. *The Morning Chronicle* found fault with a stuffed tiger's tongue.[29] And in at least four issues of the *Illustrated London News*,

---

26 Charles Waterton knew that it was the natural tendency of all cured flesh to shrink; hence, specimens were often "withered, distorted, and too small." He complained that "the feathers begin to drop off, and you have the hideous spectacle of death in ragged plumage" (325). Waterton (1782–1865) was a natural historian who traveled several times to South America. He was a highly skilled taxidermist who sometimes displayed an eccentric sense of humor by creating satirical tableaus of animals dressed as famous people.

27 After Frank Buckland's monkey "Jacko" died from bronchitis (Buckland had taken this monkey to Oxford with him and had been dismayed when it tore his notebooks), rather than turn him into a "hideous" stuffed monkey, he made his skin into a tablemat and mounted his skeleton (Buckland 319). Other taxidermists also registered their discontent with the art's inaccuracies. Montagu Browne in his *Artistic & Scientific Taxidermy and Modelling* criticized an 1892 study of an owl exhibited at the Royal Academy: "the painting was simply perfect and learned—but the bird! A ragged moth-eaten old specimen stuffed by some incompetent hand" (12).

28 Because of professional jealousies and because of the aura surrounding a gorilla specimen, people accused Du Chaillu of never actually seeing a gorilla. Bartlett's *Life Among Wild Beasts* also wonders about Du Chaillu's integrity. Referring to the gorilla in question, he recalls: "I, however, then and there convinced him that the blackness of the face was due to its having been painted black; finding I had detected what had been done, he at once admitted that he did paint it at the time he exhibited it in New York." Bartlett continues: "The question that arose in my mind upon making this discovery was, did M. Du Chaillu kill the Gorilla and skin and preserve it? If so, he must recollect that the epidermis came off; supposing he did forget this, he must have been afterwards reminded of the fact when he had to paint the face to represent its natural condition. These facts (to which I had a witness) led me to doubt the truthfulness of M. Du Chaillu's statement, and it occurred to me that he was not aware of the state of the skin, and probably had not prepared it himself" (*Life Among Wild Beasts* 254).

29 A review of "The Colonial and Indian Exhibition," in the *Morning Post* (April 22, 1886) complains: "the tigers nearest the entrance were ill-managed about the heads … the tongues, thickly painted and exhibiting no papillae, being apparently made of slabs of some material, probably of clay … ."

"SI NON E VERO," &c.

Old Lady. "Oh, Mr. Hackles, you 've stuffed my Parrot very badly! All the Feathers are coming out already!"

Taxidermist. "Why, Lor' bless yer, Mum, that 's the puffection o' Stuffin'! You know the Moultin' Season 's now a comin' on, Mum!!"

Figure 3.7 "Si Non E Vero Etc."
Source: Punch 1881.

Waterton ardently campaigned against a piece of displayed taxidermy, which to his mind was no more than a grotesquely distorted, repulsive caricature of the real thing. In particular he blasted a stuffed peacock's shriveled legs and toes, featured at the 1851 Great Exhibition. Waterton, indeed, believed that taxidermists should be trained to observe as carefully as an artist. Addressing his critics, he exclaimed: "Were you … to pay as much attention to birds as the sculptor does to the human frame, you would immediately see, on entering a museum, that the specimens are not well done" (October 11, 1851: 14).

The reason why Victorians became outraged over a badly prepared and stuffed skin was because, as I have already suggested, so much—name, character, memory—depended upon its colors, markings, and texture. As a result, pet owners wanted their pets to look as they had when alive, but most of all, as discussed in the chapter on collecting, scientists relied upon skins, whether stuffed or not, in order to progress with their efforts to classify species. While field observations were crucial to distinguishing new species, and although scientists were increasingly examining what lay beneath the skin, the ultimate truth of animal order for many still rested on the surface as well as on the existence of at least one preserved specimen, a physical proof that was available for repeated observation and investigation. Collected in museums or private homes, these specimens functioned as works of reference. It was crucial to have a specimen as close to nature as possible.

Because of the importance of accuracy, natural historians were intent upon correcting false information perpetuated, for instance, by the stuffed walrus, still on display at the Horniman Museum (London) and first exhibited at the Colonial and India Exhibition in 1886.[30] In this case, because the taxidermist had never actually seen the live animal (the specimen itself had been brought back by the Victorian hunter James Henry Hubbard from Canada), he smoothed out its skin's distinguishing, idiosyncratic folds and wrinkles, and produced a false and "overstuffed" image.[31]

---

[30] The Horniman Museum is in South Forest, South London. The Museum was commissioned in 1898 and completed in 1901.

[31] By extension, natural historians were upset that established or academy artists were perpetuating errors by using badly mounted specimens as models for their images of animals or birds; as Dr. Livingstone remarked, "painters generally make the lions' faces like old-women in night-caps" (Nott v). After attending an exhibit at the Royal Academy, Frank Buckland, one of the most committed of the Victorian naturalists, wrote that Painting No. 20 showed flying gulls over a heavy breaking sea, but he could not "tell of what species they are." Annoyed, he added, "There are plenty of cormorants and gulls at the Zoological, which would have done for models" (Bompas 267). Buckland also was outraged when he glanced at painting No. 50, "Fox Cubs": "Three heads peeping out of a hollow tree, but I wonder what they are standing upon inside the tree. Foxes' faces, especially cubs, are more difficult to paint than those of babies, the expressions of both are so varied. I never yet saw a fox's head stuffed properly, for the taxidermists generally put in *round* pupils to the eyes, like the eyes of dogs, not slit-like pupils like those in a cat's eyes" (Bompas 268). Some might have regarded Buckland's criticism to be an extension of his eccentric self, but it was not; indeed, it was mainstream, for the established journal *The Art Journal* in its August 1882 review of "Animal Painting at the Royal Academy" grumpily pointed out that in the painting "A Race for Life" the hooded crows pursuing a hare were ridiculous. They depicted "a stumpy or tail-less species not known to us" (254). The reviewer also protested:

> It does not seem to occur to men who paint such pictures as these, that it would be well before starting to make themselves really acquainted with the bird's anatomy, and the way the feathers are massed over the various muscles of the body. They never, therefore, realize that the feathers always be in certain well-marked masses, corresponding in some degree to the muscles which they cover,

For the most part, this desire for accuracy concentrated upon the specimen's surface: the colors and contours of the animal's skin or the bird's feathers. Scientists and collectors, especially, were anxious that both taxidermists and illustrators—the intermediaries between the original and its image—exactly replicate the way the bird or animal looked in nature.[32] They made a point of having these artists and craftsmen visit zoos/menageries or see the animal in the wild to record a specimen's exact coloring. Taxidermists and illustrators carried boxes of watercolors so as to make notes of the skin's colors before the feathers, fur, and skin faded after the specimen was shot. Among Lord Derby's letters is one, written in 1837, from a man in Aberdeen who, when sending a kiwi skin to the Earl, enclosed a sketch drawn from the living bird in New Zealand (at this point there was no live specimen in England). The correspondent explained that the drawing "may help with modeling the skin" (920 DER [13] 1/151/1A, National Museums Liverpool). Examples of such attention are plentiful. For instance, once Audubon had shot a bird in the field, he often nailed it to a tree in order to make a sketch on the spot and note the colors before the skin faded; John Gilbert, who worked for John Gould, tirelessly made notes on what he had just killed; and Lear spent endless hours not only sketching birds from life but also recording, in the margins of that sketch, the correct lay and colors of the feathers. The exoticism of these alien species was carried by these particulars in a manner similar to the way ethnographic portraits carefully attended to details of costume, personal appearance, and race. The focus on the details of a specimen's outer covering in natural history illustration is no different.

As the letter concerning the kiwi indicates, Lord Derby was among those who relied upon the truthful portrayal of skin if he were to realize his ambitions as a collector. In order to become familiar with and be able to recognize or name the new specimens that had come to his attention, Lord Derby depended heavily not only upon stuffed specimens but even more so upon natural history illustrations which correctly depicted a creature's outward markings. He purchased and commissioned original paintings, collected prints and watercolor sketches, and

---

> and that the feathers of the wings are always in fixed numbers, varying slightly in different species, and lying one over the other with exact regularity; had they studied the subject they would never paint these parodies of birds. (254)

For Buckland and many like him, this transparent scientific likeness was primary.

32 Sometimes a completely accurate illustration was impossible. Robert McCracken Peck's remarks about natural history illustrators who prepared prints for constant reference among collectors speaks, however, of some speculation concerning the subject's actual appearance: "It was the artist's responsibility to bring them [the natural history subjects] back to life … by fleshing out their emaciated corpses, reconstructing any missing parts, imagining the color of such fugitive parts as eyes and areas of exposed skin which change quickly after death, and, based on careful observations of related species, reanimating the subject with a typical posture, gesture, or pose" (19).

Figure 3.8 Drawing of Kiwi, 1839
Source: Courtesy of National Museums Liverpool. 920 DER (13) 1/151/A.

accumulated now rare hand-colored illustrated books for reference.[33] Educating himself so that he might recognize a species, during his earlier years he even copied some of these drawings. Later he commissioned artists, such as: Benjamin Waterhouse Hawkins; Joseph Wolf, who spent two months at Knowsley and was known, like Lear, for placing special importance on the accurate and sensitive delineation of the surface of a bird; Richard Ansdell, an artist from Liverpool who painted a special portrait of "Old Billy," Lord Derby's favorite Red Deer stag;[34] and, as already mentioned, he hired Lear to record the exterior appearance of these creatures and fasten their identity.[35] For instance, before he sent a small and very rare water buffalo, which had lived in the menagerie, to the taxidermist, Lord Derby made sure that an artist "properly" recorded its appearance on the day the animal died (Fisher 85). The illustration stabilized the instability of skin,

33 For an essay on the 13th Earl of Derby's collection see Clemency Fisher and Christine E. Jackson's essay "The 13th Earl of Derby as a Scientist" in *A Passion for Natural History* (Fisher 45–50).

34 To learn more about "Old Billy," see Fisher 91.

35 To keep an accurate record of the animals and birds in his menagerie, in 1850 the Earl of Derby privately printed his *Gleanings from the Menagerie and Aviary at Knowsley Hall*. In the "Preface" to the volume, John Gray wrote: "The following plates are selected from the series of Drawings of Ungulated Quadrupeds made by Mr. Waterhouse Hawkins for the Right Honourable the Earl of Derby, chiefly from the animals living in his Lordship's Menagerie at Knowsley Hall. They have been lithographed by Mr. W. Hawkins, and coloured, or printed in colours, under his superintendence" ("Preface" 1850). Earlier, Lear had done all the illustrations for an 1846 edition of *Gleanings from the Menagerie and Aviary at Knowsley Hall*.

the crucial medium to life, which could shrivel up and lose color, texture, and shape. The resulting drawings are in the Library at Knowsley Hall and so too is Lord Derby's extensive collection of natural history drawings and paintings. They are perfect examples of Woodall's point that "realism and truth" are "central to nineteenth-century portraiture" (5).

Because scientists and serious amateurs were often dependent upon illustrations, the London Zoological Society and Gardens, as well as many other zoos, commissioned watercolors of its most interesting specimens for its publications, and, from time to time, mounted exhibits of these portraits for the public to view. (For instance, in 1865 a gallery at the London Zoo was devoted to Joseph Wolf's watercolor drawings.) The Zoo also illustrated its "Proceedings" and "Transactions" with commissioned watercolors of its captives. And notably, John Ruskin, in his attempts to educate the people of Sheffield, hung illustrations of birds and animals by Gould, Audubon, and Lear in the rooms of his Museum.[36] For most scientists and collectors, photography, when it became an option in the 1850s, was initially a poor substitute for a drawing or a painting. Photographs did not necessarily guarantee what Woodall in her study of portraiture considers "an inherent, objective, visual relationship between the image and the living world" (6). To begin with a photograph did not capture color; it did not replicate texture, nor did it allow the viewer minutely to understand the lay of a bird's plumage (if one looks at older photographs, the exterior of a bird is blurred, approximate).[37] It was only later that people preferred to use photographs because of the medium's increasing ability to record movement as well as microscopic detail.

Part Two: Edward Lear and Natural History Illustration

When Lear was at Knowsley Hall, or even earlier at the London Zoological Gardens, and working for Gould, he would have examined numerous examples of these natural history illustrations and possibly watched taxidermists at work.[38]

---

36 For a virtual tour of Ruskin's Walkley Museum, Sheffield, go to: http://www.ruskinatwalkley.org.

37 John C. Edwards in "The Value of Old Photographs of Zoological Collections" wonders: "we must ask why nineteenth-century zoos made so very little use of photography. The first photographs ever taken in a zoo were almost certainly those taken by the Count of Montizón at the London Zoo in the summer of 1852. The photos were exhibited at the Royal Society of Arts in December of that year where they were admired by Queen Victoria. Montizón's reason for working at the London Zoo seems to have been to display his skill as a photographer, rather than to record the appearance of the animals" (Hoage and Deiss 145).

38 Lear was commissioned by Thomas Bell, dental surgeon at Guy's Hospital and professor of Zoology at King's College, London to do the illustrations for *The Zoology of the Voyage of H.M.S. Beadle*, for Sir William Jardine and Prideaux John Selby's *Illustrations of Ornithology* as well as Jardine's *Naturalists Library*, and for John and Elizabeth Gould's

Conscious of the criticism of poorly mounted and represented skins as well as the defective drawings based upon them, Lear paid extraordinary attention to the skins or surfaces of his subjects. He was well aware of how damaging an erroneous image could be to a correct understanding of a particular species. He belonged to a culture that translated surface into cultural knowledge. For instance, during his early association with Knowsley, Lear sent Lord Derby a drawing of an "Emys Ornata" [?]—Lear's script is difficult to read here. Because the creature had recently died, Lear had not been able to render a sketch from life, but had been required to borrow an earlier illustration. The letter's apology for this fact significantly reveals not only Lear's anxiety that the illustration might be flawed but also his acute awareness of the requirement that this portrait be true to life:

> I have taken the liberty of sending the accompanying drawing of which I beg your Lordship's acceptance. It is a coloured lithograph of Emys Ornata/young/ a specimen which your Lordship wished me to figure from a specimen then living at the Gardens [the London Zoological Gardens]:—since, however, the animal died before I was able to sketch it, I thought it might please your Lordship to have a drawing of it—though only in Lithography, so I have accordingly procured the accompanying plate from Mr Bell's work. ([64 C/1], November 27, 1833, The Derby Collection, Knowsley Hall)

Although Lear had no aspirations in the field of science and sometimes grew impatient with the squabbles over the identification of skins, his willingness to attend to the minutest details of an animal's hide or a bird's feathers supported the study of natural history.[39] Though he never practiced taxidermy (or even learned the art), he went to extraordinary lengths through his watercolor drawings to render an exact replica of his subject. Beginning with his early work as a medical illustrator, rendering morbid disease drawings for hospitals and surgeons, Lear was sensitive to the need to represent skin (and, therefore, the manifestation of a disease) as accurately as possible if identification or diagnosis were to be possible.[40] His natural history sketches display the same sort of diligence and commitment.

Birds of Europe. He labored on these various commissions as well as on his own portfolio *Sketches of Animals in the Zoological Gardens* and his privately printed *Illustrations of the Family of Psittacidae, or Parrots* (begun 1829).

[39] "Unlike Gould, Lear had no credentials or even aspirations in the field of science. With the help of others, he did his best to identify the birds he painted with common and scientific names, but even these sometimes proved inaccurate because so little was known about the birds he was depicting" (Peck 32).

[40] "I began to draw for bread and cheese, about 1827," he recalled late in life, "but only did uncommon queer shop-sketches—selling them for a price varying from ninepence to four shillings: colouring prints, screens, fans; awhile making morbid disease drawings for hospitals and certain doctors of physic" (as quoted in Peck 3).

On December 17, 1834, Lear wrote to Lord Derby about his efforts to capture an exact likeness of the manges oppossum's fur:

> My Lord, As I understand your Lordship is not expected in town until March, I have sent down the drawing of Manges Opossum, the skins of British mammalia [*sic*] I did not think there was any necessity for sending—& I have left them at your Lordship's house in Grosvenor Square. I hope the drawing of the Opossum will give you pleasure: I have taken a great deal of pains with it, & from having worked at Opossum a good deal lately, I trust your Lordship will think that I [have] been able to imitate the fur more nearly. ([64 C/2], December 17, 1834, The Derby Collection, Knowsley Hall)

Painfully wanting to capture the most precise image of an animal's or bird's skin as possible, Lear preferred to use living specimens rather than rely upon badly stuffed examples or flawed illustrations. When he could, Lear observed those specimens, which were still fluttering in their cages or trotting around Lord Derby's estate.[41] If none was available, he worked carefully from reliably mounted skins. For instance, he executed his watercolor of Lord Derby's woolly opossum, dated March 18, 1834, from a credible specimen,[42] and in August 1836 painted a watercolor of a woodchuck from a stuffed skin displayed in the Knowsley Menagerie Museum (Fisher 112).

The color notations as well as penciled notes in Lear's preliminary sketches are but one indication of his assiduous attention to the almost invisible, yet telling, details of a creature's exterior or surface. Surrounding his study of the rock hyrax, for example, are observations concerning the way this animal's hair falls "softly over the toes," as well as comments on the appearance of the rough ochre hair around its mouth, the "fringe of browner ochre lines" above the toes on the right leg, and the creases on the creature's paw.[43] Similar marginalia inform other studies. Lear's 1831 preparatory sketch of the red and yellow macaw, for example, displays dabs of possible colors, done in pastels or vivid hues—all dedicated to representing the appearance, especially its feathers, as accurately as possible. The result of such attentiveness is that Lear's natural history illustrations, such as "The Spectacled Owl," almost offer the viewer an exaggerated version of the real thing.[44] Because the image stands verbatim before its viewer, there can be no confusion, as there had

---

41 Sometimes he required the assistance of an interested keeper to help him keep a bird still so he could measure its various parts.

42 Clemency Fisher states: "There is no evidence that the animal was ever alive in the Knowsley Menagerie" (130).

43 For a reproduction of this sketch, see *Harvard Library Bulletin* 22.2–3 (Summer–Fall 2011): 133, fig. 34. Richard Owen did an autopsy on the animal after it died. Before this animal died, it was considered to be of sufficient rarity and so was exhibited at the London Zoological Gardens from 1828 to 1863.

44 For a reproduction of "The Spectacled Owl," see Susan Hyman's *Edward Lear's Birds* 83.

been for *Punch*'s "Old Mr. J—N—S" between a stuffed cat and a live owl (see Figure 2.6). The studied density of Lear's magnified precision allowed for no such error.

Under the Skin

Significantly, however, Lear's extraordinarily precise rendition of a bird's feathers or a mammal's fur in these portraits is not what necessarily distinguishes his watercolors and prints from those executed by other illustrators, who often worked as diligently as he to perfect these details. Rather, what makes Lear's illustrations valuable, if not endearing, is the fact that through his subject's posture and eyes, he permits, from time to time, a subtle glimpse of something vibrantly alive "under the skin" or inherent within the dense details of the subject's fur or feathers. An attitude or an expression of individuality comes through the opaque density of these minutely rendered particulars and removes the subject from being simply an exemplum of its species. No longer is it "a mere piece of property" (Freeland 32)—as the children's rhinoceros becomes in "The Story of the Four Little Children." For instance, Lear's "The Red and Yellow Macaw" defiantly turns its head toward the observer (one almost sees the twisting of its neck), as if responding to the artist's gaze; the pleading direct glance of the "Bay-headed parrot," from *Illustrations of the Family of Psittacidae, or Parrots*, and the mocking glance of the "Spectacled Owl" also engage the viewer, and suggest that more is at play than the carefully rendered lay of their feathers. In these illustrations, and many like them, Lear and his subject seem to be conversing with one another; moreover, his subject seems to be consciously posing. Its position is not one superimposed, so to speak, by the portraitist. In this manner, these watercolors/prints compromise, if not defy, the conventional animal portraits in which there is no looking back and, therefore, no negotiation. No longer is the subject present only to be stared at.[45]

In this sense, Lear's natural history illustrations challenge Cynthia Freeland's assertion that it is impossible to render portraits of animals. In her *Portraits and Persons*, she asks the question: "Can there be portraits of animals?" and answers in the negative because, from her perspective, "portraiture involves an act of posing or self-representation" (17). I suggest, however, that in Lear's work, animals seem almost to pose or bargain with the artist concerning how they will be represented. A transaction between the artist and animal becomes a possibility, for both bird and artist do appear mutually to be conscious of one another. They give the impression of being equally involved. Lear's portraits of animals and birds are truly proof of contact.

Because of this "contact," it is no wonder that when John E. Gray, curator of the British Museum, wrote to Lord Derby concerning the proposed publication of *Gleanings From the Menagerie and Aviary at Knowsley Hall*. Gray adamantly favored Lear's watercolors, for their vitality and grace as well as for their "very accurate representations of living specimens" ("Preface" 1846 edition). He wrote

---

[45] Although it is often written that Joseph Wolf was the most distinguished of the bird illustrators, his drawings rarely intimate an existence beneath the skin.

to Lord Derby: "I have been considering with care the drawings you have sent up to me. The Drawings of Lear are in such a different style, & so superior in artistic style to those of Mr. Hawkins [Benjamin Waterhouse Hawkins] that I think they should hardly be mixed with his … 30 very useful & beautiful Plates might be selected from Lears [*sic*], which would be most useful to Scientific Man" (December 14, 1840, National Museums Liverpool).

Part Three: The Reversals and Revenge of Nonsense

When one turns to Lear's nonsense drawings after glancing at his commissioned illustrations, the subjective and implied communion between observer and creature becomes even more explicit. In his limericks, for instance, the Raven and the Old Man of Whitehaven literally exchange delightful glances, imitate each other's movements, and dance in synchronized rhythm; similarly, the Old Man and the owl imbibe ale together while sitting on a fence in a parallel posture, as if animal and human have spontaneously, but consciously, negotiated some sort of understanding. Released from the restrictions and expectations of convention associated with natural history illustration, in these nonsense sketches Lear revives the subjective life inherent in the skin of an animal or a bird, and, thereby, rebuffs the colonial's and the collector's commanding gaze upon the skin's surface. Through his nonsense, Lear momentarily liberates portraiture as well as his subjects from the prerogatives of classification, ownership, and commodity.

The Nonsense

From time to time critics have recognized that Lear's nonsense drawings, especially those initially written or sketched to entertain Lord Derby's extended family, are indebted to his immersion in the Knowsley culture and his professional commitment to natural history illustration. Clemency Fisher, for instance, comments that in the cartoon accompanying "The Pelican Chorus," both the pelican and the composite bird "with the upper half Blue Heron and lower half Stanley Crane" were based upon live specimens inhabiting the Knowsley Menagerie (120).[46] Inevitably, his other work as a naturalist crept into his nonsense. It is therefore not unusual to see the storks, parrots, and owls of his studies transformed into nonsensical caricatures. The "Black Stork" which Lear rendered for Gould's *The Birds of Europe* (V.4) finds its way into the illustrations accompanying "There was an Old Man of Dumblane" and eventually into *A History of Seven Families of the Lake Pipple-Popple*. These transformations are often acknowledged. What is not necessarily documented or discussed, however, is that when dashing off his nonsense drawings, Lear turned

[46] A cursory glance through *A Book of Nonsense* and its various sequels reveals the cranes, the owls, the pelicans, the apes, the zebras, and the crocodiles, which were all part of the menagerie culture surrounding Lear.

his natural history assignments upside down in order to empty the commissioned illustrations of the suffocating details that take the life out of the portrait. His subjects can once more breathe and more fully realize the life residing within their skins. In a sense, his nonsense *unstuffs* the specimens standing either in the museums, preserved by taxidermy, or hanging on the walls of the Zoological Society.

When drawing his nonsense, Lear showed absolutely no interest in including the painstaking details of his commissioned work—those surface particulars that scientists and collectors felt were necessary to name, identify, classify, own, and control a species. In his nonsense sketches, Lear deliberately emptied his meticulous studies of birds and animals of every vigilantly rendered feather or mark, and left merely an effortless fluid outline of his subject. Through these inversions, Lear created a certain spontaneity, which he understandably could not afford to let dominate his commissioned pieces. To experience this difference, compare, for instance, the watercolor of his painstakingly detailed "Spectacled Owl" with the hastily and casually sketched owls sitting in the nest with the depressing "Old Person of Crowle," and place the same natural history illustration either next to the owls perched on a railing while being taught to drink tea ("There was an Old Man of Dumbree") or beside the owl sitting on the fence with the old man ("There was an Old Man with an Owl").

Figure 3.9 Lear's "Old Person of Crowle"
Source: More Nonsense, Pictures, Rhymes, Botany, Etc. 1872.

Metamorphosed into nonsense, the "Spectacled Owl," which had formerly filled the illustrated page with the density of its skin's particulars, is now reduced to an inscribed outline. The carefully rendered details have receded into an approximation and gathered into simplified lines. In a way, the resulting inscribed image suggests the indelible impression left just underneath the surface of Lear's own psyche or buried within the skin by the intense demands of his immersion in the natural history culture. Like the tattoos once worn by the sailors, on display at the Wellcome Institute, these cartoons register the lasting

Figure 3.10 Lear's "There was an Old Man of Dumbree"
Source: More Nonsense, Pictures, Rhymes, Botany, Etc. 1872.

Figure 3.11 Lear's "There was an Old Man with an Owl"
Source: The Book of Nonsense. 1915.

influence of time, place, and work. They carry a memory of the labor associated with pieces such as the "Spectacled Owl," in a manner which reminds one, admittedly in a rather tentative way, of the permanent blue lines once staining the skins of coal miners. These lines were witnesses to these men's labor. In *The Road to Wigan Pier* George Orwell talks about the coal dust, which lodged in the creases of the miners' skin, and left indelible lines no matter how often these people scrubbed their faces and backs.

The Reversals

Released from the densely particularized envelope which scientists and collectors relied upon to define and limit their identity, the creatures frolic on a frameless page and reveal the energy dormant within their skins.[47] Liberated from the required and suffocating details, their skins can once more breathe; consequently, Lear's natural history subjects are now able to dance a quadrille, to leap, to rebel against being "civilized" by being taught to drink tea, to seize a man's nose, or to sit on an old man's nose as well as a young lady's bonnet. No longer are the owls, the parrots, the zebras, the ravens, and the Barbary apes captives of the critical imperial gaze. (I like to think that Lear, through his nonsense, could have actualized the dormant life inherent in the stuffed Zebra's skin that I saw in the Zoology Museum at Cambridge University.) No longer are they static commodities, official representatives of a species, passively waiting to be stared at or studied; instead, they themselves are the ones who look and move as conscious, liberated beings. In "There was an Old Person of Crowle," for example, all the owls turn to look disapprovingly at the depressing old man who sits in their nest (and screams out with the rest). The colonial gaze is reversed; the tables are turned. Instead of being the victim of such scrutiny, the bird, ridiculing the imperialistic notions of observation, now stares back and gravely focuses on the person. Lear's drawings consciously react to the reality which John Berger was much later to recognize: that "Animals are always observed … They are the object of our ever-extending knowledge. What we know about them is an index of our power, and thus an index of what separates us from them" (Berger 14).

With the loss of the imperialistic gaze, Lear also topples the colonial's mistaken sense of his own importance and superiority. He mocks what he was seeing while working with ornithologists, zoologists, collectors, keepers, hunters, breeders, and exhibitors. He turns their myth of superiority upside down. For instance, the Old Man who said "Hush!" looks at a bird and has no choice but unconsciously to mirror it. He is not its better; he is just like what he is observing. Indeed, he sits in the nest with the bird. Similarly, the old man of Dover looks like the bees which pursue him. He cannot outstep either their image or their sting; he is the one preyed upon. And the old person of Nice who associates with geese as well as the old man of El Hums who lives on nothing but crumbs must suffer the "tolerant" or amused glances of the birds these so-called superior figures unknowingly imitate. As always with Lear, the birds possess the controlling eye. They, like the bird in the bush, stare right back in a manner that recalls a preparatory pencil sketch of an unidentified parrot in which the bird gazes disapprovingly at a stout man who stands face to face before the bird.

---

[47] Lear seems to have tucked his nonsense into parts of the day not connected to the labor accompanying sunlight; he often squeezed it into the corners of his correspondence and of the pages of his diaries. (Colley, "Edward Lear's Limericks and the Reversals of Nonsense" 286.)

Figure 3.12 Sketch of unidentified parrot by Edward Lear
Source: Houghton Library, Harvard University.

In a sense these reversals, which put the colonial and collector "in their place," express Lear's impatience and discomfort with the imperative to colonize and order the animal world. Though Lear was friends with those responsible for collecting, organizing, and enclosing exotic species, and even enamored with or dependent upon them himself, he stood on the boundaries of that culture and longed for a time when creatures were not colonized or objectified but, instead, enjoyed a more harmonious, symbiotic relationship with humankind.[48] Feeling marginalized (did he belong at the housekeeper's or his Lordship's table?) by those who had advantages as well as by his epilepsy, not to mention his sense

48 In addition to his dependency upon people such as the Earl of Derby, throughout his life Lear's acquaintances and closest friends (significantly, they were also his patrons who supported his painting and travels) were those administering the various British protectorates. Clearly Lear was capable of subscribing to the colonial system and its prejudices against the non-English. For instance when feeling uncomfortable, he could register his disgust with such figures as "a big, horrid, vulgar, ill-dressed, gross blacky Indian" or speak of "filthy Arab savages" (Colley, "Edward Lear's Anti-Colonial Bestiary" 110).

that people, according to his own words, found him "ill-tempered and queer" ("How Pleasant to Know Mr. Lear"), Lear consequently discovered a pleasant and relatively safe companionship with animals. His celebrated attachment to Foss, his cat, is an example. The numerous drawings of him and Foss stepping in an ironic union are, perhaps, testimony enough.

He identified with animals. Often in his self-deprecatory portraits, he depicted himself as a rather stout bird with stubby wings. On other occasions he used animals to portray his state of mind. In the late 1830s, for example, when

Figure 3.13 Sketch of Phos By Edward Lear
Source: Houghton Library, Harvard University.

he arrived in Italy to study landscape painting, he exclaimed, "I am extremely happy—as the hedgehog said when he rolled himself through a thistlebrush" (Noakes 58). In addition, Lear was fond of referring to himself as an animal. When he was working on his parrot studies, he proclaimed, "for the last 12 months I have so moved—thought—looked at,—& existed among Parrots—that should any transmigration take place at my decease I am sure my soul would be very uncomfortable if anything but one of the Psittacidae" (Noakes 33). Ideally Lear would have preferred a more fraternal partnership between man and animal. His drawings of the old man of the Border dancing with his cat or the old man in the tree who contentedly allows the birds to pluck his long hair for their nests are part of this desired symbiosis in which man and animal are in harmony.

Figure 3.14 Edward Lear's "There was an Old Man on the Border"

Moreover, as his nonsense obviously testifies, Lear was sometimes uncomfortable with the scrutiny, classification, and implicit authority to which a creature was subject, so he ridiculed imperialistic notions of ownership and observation. He enjoyed poking fun at the rage to classify and, thus, control the natural world. His nonsense botany, in which, for instance, he identifies the "Manypeeplia Upsidownia" and draws people like petals hanging upside down on a stem, as well as his sets of Coloured Birds ("The Dark Blue Bird," "The Pink Bird," and "The Light Green Bird"), illustrated for a child he knew in Corfu (Mary de Vere), are examples of how he relished unsettling the order of things.[49]

[49] One of the most endearing pieces of zoology is his "Portraites of the inditchenous beestes of New Olland" (1838), which shows Lear's humorous sketches of animals he and John Gould saw in Rotterdam, Berne, Berlin, and Amsterdam, either in 1828 or 1830.

Conclusion

Clearly, "The Story of the Four Little Children Who Went Round the World," discussed in the chapter's opening is one instance of this critique of such colonial practices. The rhinoceros's fate is a reflection of Lear's discomfort that in the end an exotic creature is nothing but a utilitarian skin, figuratively transformed into a doorstop and made into an emblem of England's dominion over remote territories and nations. The story represents Lear's own ambiguity when sketching at the London Zoological Gardens, accompanying John Gould on his visits to zoos, and studying the animals inhabiting the Knowsley Hall Menagerie. Uncomfortable with the scientific and imperialistic practices of caging, classifying, and turning these animals into commodities, in his illustrations he yearned to revive the life residing within his subjects' skins by ridding them of the suffocating details required by his commissions. He wanted to restore their intimacy and character. And, as in the case of the rhinoceros as well as other taxidermy specimens, he desired to take out the stuffing and let the skin once more recover the texture of its life and breath. In his nonsense, he makes explicit his desire to create another kind of portrait from those done to satisfy collectors and scientists—portraits that do not depend upon the stranglehold of particulars but, rather, acknowledge the intimacy of life inherent in skin.

Chapter 4
Touch: Reaching through the Bars

To make the sight as true as touch
William Hazlitt, "The Indian Jugglers"

Introduction: Reaching through the Bars

Nineteenth-century descriptions of Victorian British zoological gardens and menageries are remarkable for their multiple accounts depicting the public's eagerness to reach through the bars or wire mesh of cages so as to be able to touch or "caress" the enclosed creatures. (These nineteenth-century texts are fond of choosing the word "caress" to portray this activity.) Desiring to stroke, pat, embrace, and fondle these exotic animals, children and adults introduced their hands into the dens so as to sense the animal's hide brushing up against their fingers or to feel the lick of the creature's tongue.[1] The 1829 *Picturesque Guide through the Regent's Park*, for instance, describes three young wolves who "have been seen to lick the hand of a visitor and fawn like dogs" (50), and *The Pictorial Museum of Animated Nature* (1843–1845) mentions a lion seen "licking a man's hand" (1: 24). Drawings and photographs from the period frequently capture these encounters—indeed, it is striking just how often these moments are featured in nineteenth-century pictorial representations of zoos.[2] One turn-of-the-century postcard, for instance, shows a person reaching his or her hand through the bars in order to have contact with a rhinoceros. In this image, the reaching hand is as prominent as the animal's open jaws. (One wonders what will happen next.)

---

[1] The impulse to touch the wild is still alive and well. Recently while visiting the Royal Ontario Museum in Toronto, I watched and listened to adults and children asking to "touch" a live tawny owl on display in a special exhibit. Longingly, the public extended their hands toward this owl, which was attached to a keeper's wrist. I also participated in this desire when, on a trip to South Africa, I stood somewhat self-consciously, yet excitedly, in a queue of tourists so that I could place my hands on the back of a cheetah, stroke it, and have my photograph taken while doing this. In this respect, it is also interesting to be reminded of a July 10, 2013 article in *The New York Times*, "Step Right Up, Kids, the Tiger Will Look Good in Your Photo," in which the reporter, Andrew E. Kramer, talks about the Nikulin Circus in Moscow where children are photographed (with their parents' permission) sitting right up next to and with their hands on one of the circus's tigers.

[2] Few modern keepers can suppress a shudder when they are shown photographs of children embracing, for instance, a sub-adult male chimpanzee in the children's zoo at the London Zoo. These keepers are well aware of the fact that this animal is capable of ripping them limb from limb.

Plate 1 "Rhinoceros, Zoological Gardens, London"

Figure 4.1 "Child extending hand through bars of rhinoceros enclosure"
Source: Nott 1886.

A photograph taken around the same time shows a child's hand nuzzling a rhinoceros, and in yet another picture (this time an earlier nineteenth-century drawing), a woman bends down so that she might put her fingers through the beaver's enclosure and stroke the creature's fur. Within the context of the zoo, there were additional opportunities for touching when keepers took young lions, tigers, and apes for walks through the zoo grounds so that children could pet and play with them. Elephant and camel rides through the grounds also encouraged visitors to fondle yet more of the zoo's inhabitants.

Notable too are the numerous anecdotes about zoo employees who habitually stroked their exotic captives. These narratives emphasize the animal's pleasurable reaction to their keeper's touch. They speak of the bear who affectionately responded to its keeper's "caresses"; the lemur who was always pleased to be stroked; the hippo who loved to be scratched; a young lion who evinced pleasure at being repeatedly petted; the tigers who delighted in being tickled and rubbed; the giraffe who bent its head down to caress his keeper; and the hyena who "suffers himself to be caressed" (Blunt 102).[3] A drawing of "Moti [the leopard] and Its Keeper" and a photograph of an employee stroking a tapir at the London Zoological Gardens illustrate these seemingly gratifying encounters. These moments and these images vicariously register the public's desire to touch these exotics.

Figure 4.2 Beaver Enclosure
Source: Zoological Keepsake 1830.

3 There are also sentimental accounts of this touching. One is about a chimpanzee who put her arms round the director's neck, placidly kissed him three times, stretched out her hand to him, and died.

Plate 2 “Camel Ride”

Figure 4.3 “Best of Friends—Hippopotamus and Keeper”

Figure 4.4 "Moti and its Keeper"
Source: Scherren, *Walks and Talks in the Zoo* 1901

Figure 4.5 "Keeper Stroking Tapir"
Source: Nott 1886.

With all the transporting of animals, zoos were, of course, not the only venue for these encounters. Captured animals being transported on vessels from foreign ports to England (an experience already discussed in Chapter 2) also inevitably offered opportunities. As one can imagine, these captives were in an awkward and close proximity to both crew and passengers. George Catlin's narrative of his 1839 trip on board a packet-ship, the *Roscius*, from New York to Liverpool (*Adventures of the Ojibbeway and Ioway Indians in England, France, and Belgium*) contains a vividly disturbing example. Catlin had brought with him not only eight tons of crates containing Native American artifacts for an exhibit at the Egyptian Hall, Piccadilly but also two grizzly bears, which he planned to display throughout Europe. He kept these poor creatures in "a huge iron cage," above deck (1: 2). Unfortunately, the voyage was unusually turbulent. The bears had a terrible time of it; "howling," "bellowing," "growling," and "raging," they were sick, lonely, and miserable (1: 2–5). (Once placed on a train in Liverpool, they continued to howl all the way to Euston Station.) To add to their distress, these grizzlies were subject to the crew's attempts to "caress" and shake hands with them. During their leisure hours, the crew frequently entertained passengers by encouraging the bears to extend their "arms" through the bars so these travelers could finger and grasp the creatures' fur.

Not surprisingly, this contact was not always endearing. In the case of Catlin's grizzly bears the result was, at least on one occasion, disastrous. One of the bears extended her "arm" right through the enclosure's iron railings and "made a side-lick" at the sailor's head. The result was, as Catlin graphically relates, that "one of her claws carried away entirely his nose, leaving it fallen down and hanging over his mouth, suspended merely by a small piece of skin or gristle by which alone he could claim it" (1: 5). As can be expected, similar consequences also awaited those who, motivated by the desire to touch the animal's skin, intermittently reached into zoo enclosures. Loss of life, mutilated limbs, ripped clothing (there were often complaints from "ladies" who had their dresses torn in the monkey house), and lost thumbs or fingers were not an uncommon consequence for visitors to zoological parks. For instance, in August 1861, a woman lost a "thumb and finger of her right hand when bitten by a young wolf," and a nine-year-old boy was badly mauled by a leopard (Blunt 206). These calamities resulted in severe warnings cautioning visitors about the peculating propensities of the monkeys, as well as notices "respectfully" directing the public "not to touch any of the animals" (Scherren, *The Zoological Society of London* 57). Alarming headlines in the press, complete with graphic pictures, told about "frightful" occurrences in which, for instance, a visitor had been seized by a bear, an employee's arm had been mangled by a tiger (the man was removed with difficulty from the tiger's jaws), and a wolf had bitten off the arm of a little boy who "had taken much pains to introduce it through the bars" (*The Zoological Keepsake* 4). As one late Victorian commentator remarked, "The impudence of caressing wild animals often carries the punishment with it" (*Picturesque Guide through the Regent's Park with Accurate Descriptions* 56).

At times, the touching of these exotic creatures was aggressive, for, as is well documented, visitors also poked their umbrellas, sticks, and whips through the cage openings so as to excite, irritate, or catch the enclosed creature's attention. (There was even a craze of carving one's name on the rump of the rhinoceros.) As a result zoos also had repeatedly to caution the public "against irritating any of the animals, or imprudently venturing on their docility" (*Picturesque Guide through the Regent's Park* 56). At Belle Vue Zoo in Manchester, keepers had the authority "to remove at once … any person found annoying, teasing, irritating, or injuring the animals in any way" (*Guide*, Belle Vue Archive, Chetham's Library). An 1849 cartoon from *Punch* humorously criticizes this irresponsible recreational behavior. The drawing, "A Prospecte of Ye Zoological Societye: Its Gardens," not only shows the crowds pressing up against the cages but also shoving stuff through the bars, poking at the monkeys with the points of parasols, and gleefully annoying the enclosed creatures.

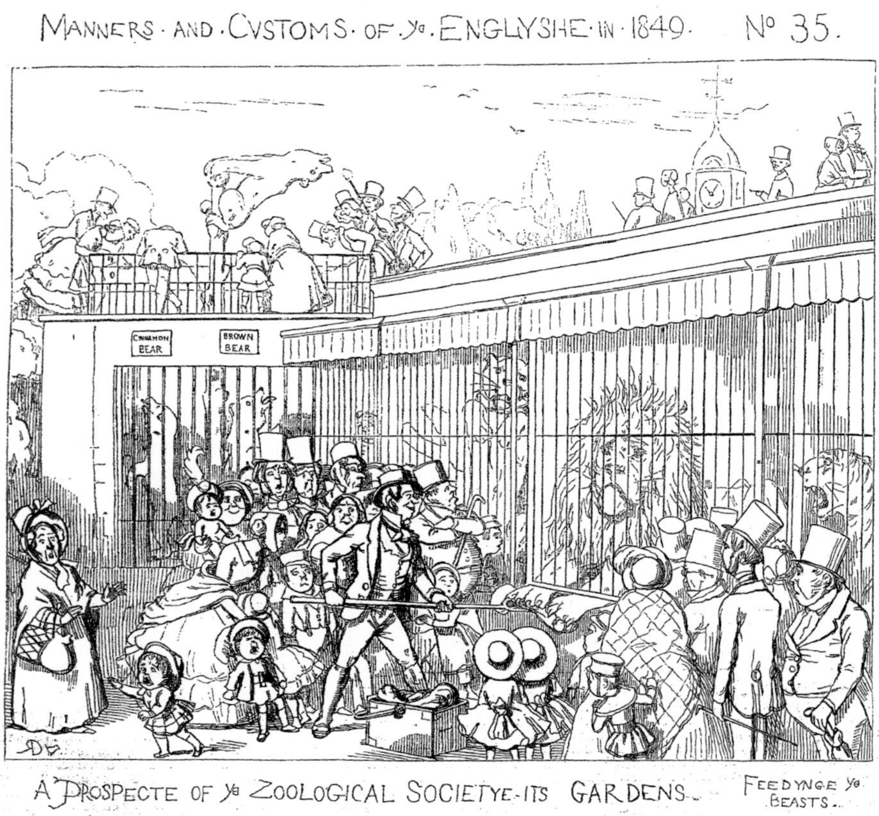

Figure 4.6 "A Prospecte of Ye Zoological Societye: Its Gardens"
Source: Punch 1849.

Needless to say, Victorian zoos and menageries allowed a great deal of interaction between the animals and their admirers. There were not always guardrails surrounding the cages, nor were there reinforced bars or additional layers of glass or wire mesh. When considering this circumstance, Harriet Ritvo suggests that zoos purposely allowed access so as to encourage people to think of their wild, exotic captives as the property of empire. From her point of view, the creatures were put on display as "temporary possessions or playthings," as "mascots," to be tamed, domesticated, or "civilized" by the ruling colonial power (*Animal Estate* 219). Although there is some truth in Ritvo's postcolonial perspective, I believe that when visitors felt moved to touch or poke these exotic animals, more than appropriation was taking place. Rather, these individuals were exercising a desire that exceeds the boundaries of a particular historical moment. The irresistible impulse to extend the hand and touch these animals, whether executed endearingly or aggressively, was not simply the consequence of a sense of some imperial privilege. Instead, the act was linked to a much deeper, and daring, desire to reach out, whatever the cost, and feel the exotic other, to go beyond the boundaries of one's own skin and actually finger the fur of a wild creature. This impulse reminds me of Steven Connor's comment that skin is "ever hungry for the touch of new impressions" (141). More is at play. I agree with other skin theorists like Penelope Deutscher, who in her essay in *Thinking Through Skin* observes that "Touching the skin of the other is *not* simply an example of the constant drive to appropriate the other" (Ahmed and Stacey 145 [italics mine]). This gesture is an expression of a larger desire to reach for what is beyond the self. It is not just motivated by the desire to take or possess.

Touch and Cognition

The desire to place a hand through the bars, to link skin to skin, is attached to the idea that touch is one of the more important aspects of knowing and that skin, as Claude Bouillon popularly observes, is "the seat of perception" (49). Skin is the organ that conveys sensations of touch, such as a light brush, contact, pressure, heat, cold, dryness, and pain. It gives us our knowledge of depth or thickness and form.[4] Those who write about the cognitive role of touch often recall Erasmus Darwin's understanding that we acquire "our tangible ideas of objects either by the simple pressure of the organ of touch against a solid body, or by moving our organ of touch along the surface of it" (*Zoonomia* 1794, as quoted in Montagu 201). A contributor to a recent study of skin, *Sk-interfaces*, revisited this perspective when

4 For an expression of this idea, see Ashley Montagu, *Touching: The Human Significance of Skin* 1. The understanding of skin as the place of perception is shared among many theorists who think about the nature of skin.

he proclaimed, "Only something we can physically grasp embodies true reality for us" (Hauser 149).[5]

With these principles in mind, one can appreciate that the Victorian zoo visitors were using touch not so much for appropriation but for knowledge. Putting their hands through the bars gave them direct access to another being, to its texture, temperature, its movements, and its pressure.[6] This immediate encounter between the skin of their fingers and the warmth of the creature's hide allowed them better to grasp the depth and thickness of the animal form as well as to achieve a more tangible idea of its spatial existence. Santanu Das's understanding, in his moving study on the role of touch in World War I poetry, that the sense of touch defines space and guides the rhythm of experience is apt.[7] In the context of the zoo, these tactile moments gave access to and verified the creature's presence and briefly linked the viewer with the exotic. Touching was essential to interacting with the foreign; the act broke down the separation between the self and the other. In this regard, when generally thinking about the experience of touch, Sara Ahmed and Jackie Stacey observe that "Skin opens our bodies to other bodies: through touch, the separation of self and other is undermined in the very intimacy or proximity of the encounter" (6). Touching animals in the zoo was not, therefore, the one-way street that Ritvo suggests. It not only gave access to another being; the experience was also mutual. It was a version of what Maurice Merleau-Ponty identifies as a "double sensation" when a person touches his own hand and alternately or simultaneously feels himself touched.[8] Merleau-Ponty's understanding that "I can feel myself touched as well and at the same time as touching" is relevant (*The Visible and the Invisible* 142). As a consequence, when the public fingered the zoo animals, they were involved in a reciprocal process. When they reached out and felt a lion's or a wolf's fur, they also brushed up against or were touched by their own wildness (the primitive side of their nature)—a rather daring gesture at a

---

[5] See Zane Berzine's "Re-Thinking Touch" in *Sk-interfaces: Exploding Borders—Creating Membranes in Art, Technology and Society* (Hauser 147–9). In talks given at the 2013 North American Victorian Studies Association Conference, Pamela K. Gilbert and William A. Cohen spoke of the ability of touch to transmit knowledge. Gilbert reminded her audience of Alexander Bain's *The Senses and the Intellect* (1855) in which Bain observes that touch was an intellectual sense like the eye and that notions of size, shape, direction, distance, and extension could all be acquired by touch. A similar paper by Marie Banfield, "Mid-Victorian Psychology and the Aesthetics of Touch" at a conference on The Victorian Tactile Imagination, Birkbeck College, July 19–20, 2013 also recalled this understanding of touch.

[6] Tiffany Field in her study *Touch* (2003) mentions that "The term touch includes several tactile senses: pressure, pain, temperature, and muscle movement" (79).

[7] For a discussion of how the hand feels, see the opening pages in Santanu Das's *Touch and Intimacy in First World War Literature*: "The sense of touch defines space and guides the rhythm … as if new eyes have opened at the tip of the fingers" (1).

[8] This frequently quoted observation can be found in Merleau-Ponty's *Phenomenology of Perception* on page 106.

time when evolutionary theories were questioning the boundaries between human and animal and were challenging the old conception that the barrier separating men from brutes is, as William Bingley the popular nineteenth-century naturalist asserted, "fixed and immutable" (as quoted in Classen, *The Deepest Sense* 109).[9]

Reaching through the bars was also connected to the desire for metamorphosis—to be something we are not—to a yearning to take on a second skin, specifically to drape the animal hide around oneself, engage its symbolic value, feel the stimulation of that skin, and reinvent the self through the physicality of the exotic touch.[10] Seeking a second skin, the zoo visitors expressed a longing to be more than themselves, to break down the barriers and confuse the distinction between animal and beast as well as home and abroad.[11] This metamorphosis made the person more complete, for the animal skin released and realized a latent quality (maybe power or sexuality), which, otherwise, would remain hidden and not be as tangible. Through the skin-to-skin contact between human and beast, energy was transferred from one body to another. One is reminded of Heracles, who wore the skin of the Nemean lion so that he might gather and retain its strength. One also recalls that in seventeenth-century Europe, Peter Paul Rubens painted his young bride's naked body, enticingly enveloped within a bear's skin (*Hélène Fourment with a Fur Coat*), so as to express what he perceived to be her sexual energy as well as his own reawakened animal desires. Furthermore, in Victorian London, Algernon Moses Marsden chose to rest his body and arm against a tiger skin draped over the back of a leather chair, in James Tissot's 1877 portrait of him, so as to borrow from this exotic beast's strength or authority. And in early twentieth-century France, the black American cabaret entertainer Josephine Baker posed and displayed her naked body on a tiger-skin rug to "remake herself in the skin of the other" (Cheng 13), as well as simultaneously to mock and adopt the lure of animalized femininity. As Anne Anlin Cheng suggests, Baker's skin was

---

9 William Bingley was an early nineteenth-century popular naturalist. For instance, in 1813, he published *Animal Biography or Popular Zoology*. Recently complicating this debate, Mary Midgley in *Animals and Why They Matter* asserts that animals and humans "are incurably members one of another" (21). However, she recognizes that "species-bonds are real" (106). She writes: "with an animal, to know the species is absolutely essential. A zoo-keeper who is told to expect an animal, and get a place ready for it, cannot even begin to do this without far more detailed information … Even members of quite similar and closely related species can have entirely different needs about temperature and water-supply, bedding, exercise-space, solitude, company and many other things" (98–9).

10 In *The Deepest Sense*, Constance Classen observes: "by putting on animal skins humans [in the premodern world] symbolically took on something of their identity" (105).

11 In *The Book of Skin*, Steven Connor notes that people seek a second skin: "You could be made more entire, more yourself, by taking on another's skin" (10). One cannot also help but recall the contemporary French performing artist ORLAN and her grafting of other skin cells to refashion her body. In "Harelequin Coat," for instance, she created a hybrid skin from her own skin cells, the skin cells of a fetus of African origin, and the muscle cells of a fat-tailed dunnot (a marsupial). Skin is a textile she wears and changes.

constantly referring to other surfaces and textures: to animals, cloth, and shadows. Aware that skin is an infinitely dressable surface, Baker was always breaking out of her own racialized skin and taking on or fingering a second skin.

To further demonstrate these qualities associated with touch as well as to offer another illustration of the desire for metamorphosis—the yearning to take on another skin—it is interesting briefly to return to George Catlin's experiences during his mid-nineteenth-century tour of Europe with his Indian artifacts. At one point in Manchester, Catlin hooked up with a traveling show of "nine wild Indians" (the Ojibbeway) (1: 103), from the backwoods of America, who were on display at various assembly rooms and exhibition halls. It is interesting to note that this group of Ojibbeway was thought to be particularly wild (closer to a wild animal than to a human), for the story had circulated that during their visit to the London Zoological Society all the animals had become excited and started to howl: one of their kind had entered the grounds. In his description of his visit to Manchester, Catlin describes the immense excitement among the crowds who lined the streets to catch a glimpse of these Ojibbeway, dressed in feathers as well as in the white skins of mountain sheep. These occasions bear a similarity to a parade of wild animals being taken to a circus or a menagerie. (The streets in the vicinity of the Indians' hotel "became so completely besieged, that a strong party of police was necessary to keep back the crowds" [1: 106].) Catlin also records the reaction of those Mancunians who were in the rooms and halls where these Ojibbeway performed. These people came not only to see "Real, Red Indians" but also to mingle with them. As if reaching through the bars erected by the usual segregating social and cultural barriers, they extend their hands so they might actually touch the Ojibbeways' faces and arms, smeared with bear grease and "streaked in vermillion" and black paint (1: 109).

No. 6.

Figure 4.7 Adventures of the Ojibbeway and Ioway Indians in England, France, and Belgium

Source: Catlin 1852.

According to Catlin, ecstatic ladies were said to crowd round the troupe so as to stroke the Indians' naked skin, and, in turn, be caressed by these wild men. These ladies, it seems, literally and figuratively brushed up against a vestige of their own wildness. They were performing another version of the act of placing a hand through the bars of a zoo enclosure so as to touch a wild skin. They were thrilled to find that the "war paint" left traces on their own faces and hands. In a sense, for a moment, these ladies took on a second, wild skin—that is, until they washed off the traces of paint (cleansed and civilized once more) with soap and water. A passage describing a gathering in an assembly hall in Manchester demonstrates the virulence and contagion of this encounter with a wild skin:

> Many ladies were offering them [the Indian troupe] their hands … some were kissing them … The women commenced it as *Sah mah* had dashed into the crowd; and as he was wending his way back, finding it had pleased so well, he took every lady's hand that was laid upon his naked arm or his shoulder as a challenge, and he said that he kissed every woman that he passed. This may or may not be true; but one thing is certain, that many there were in the room that evening who went home to their husbands and mothers with streaks of red and black paint upon their cheeks, which nothing short of soap and water could remove. (1: 68)

The Hand

So far the discussion of touch has been primarily focused not just upon skins but also upon the naked hand: the zoo visitor's eager fingers brush against the lion's skin, receive the wolf's attentions, or scratch the hippo's hide; the hands of Rubens's young bride, Hélène Fourment, clutch the bear's fur draped around her naked body; Marsden's elbow lolls upon the tiger's nape; Baker's exposed fingers rest on the tiger's skin; and, as we have just read, the Mancunian ladies' hands push through the social barriers to touch the Ojibbeways' wild skin. For most theorists of touch, the hand is the place of encounter; it is the principal organ of touch and, by extension, the seat of knowledge and perception.[12] As is commonly known, Kant casually referred to the hand as "the window of the mind" (Sennett 149). With its multiple nerve endings, the hand can explore the world, feel its pressure, temperature, movement, and texture as well as inscribe itself upon it (e.g., write one's name on the rhinoceros's rump!). As Raymond Tallis observes, the hand is what grasps, seizes, pulls, plucks, picks, pinches, presses, pats, pokes, assigns, sorts, and classifies, so that one might explore and understand the world

---

[12] According to Claude Bouillon, the skin of the hand is the most sensitive, for the pads of its fingers contain receptors—2,000 per square centimeter—which can detect several milligrams of pressure (50).

outside the boundaries of one's own body.[13] It has a "crucial cognitive role" (*Michelangelo's Finger* 22). Through it, as Diderot reminds us, one can even see in the dark.[14] As if tipped with Juno's inquiring/jealous eyes mounted in peacock feathers, each of its fingers sequentially examines the texture, space, and shape of what is exterior to it.[15] Collectively the hand's receptors survey what is within their reach and situate the body as well as the other body within that landscape. The blind man "knows quite precisely through the sense of touch what branches and leaves, or an arm and finger, are" (Merleau-Ponty, *Phenomenology of Perception* 224). The touch of the fingers is direct.

But what happens when the hand is absent? How does a person experience touch without it?

Haptic Visuality: The Absent Hand

The question: "what happens to touch when the hand is absent?" prompts thoughts about the act of gazing at paintings in which the sensation of touch enters through the eyes rather than via the direct avenue of the hand. In order to "grasp" a landscape or a portrait exhibited in a gallery, people seem to search with their eyes and not their fingers. Positioning their bodies well beyond the picture's frame, visitors to a gallery tend to stand at a distance from the work of art. Well aware that when regarding a painting they are looking at a visual imitation (and not the real thing), the removed posture of these spectators tacitly acknowledges that they cannot literally enter the space of the representation. (One cannot actually step into the room of a Vermeer painting and brush against the oak table within it.) Stepping back, viewers conform to their marginality and seem to give primacy to vision. Actual touch seems remote, if not impossible. The flat, two-dimensional canvas seems to proffer no inviting openings, such as those offered by the gaps between the bars of the zoo's enclosures or by the intermittent spaces within the crowded assembly hall in Manchester. In galleries and museums, especially, such contact is taboo and taken far more seriously than the warnings once posted in Victorian zoos.

This condition of "Don't Touch" is now sufficiently ingrained so that even if there is an impulse to place the fingers on the exhibited work, people do not dare (unless their intention is to destroy the work of art). Gone are the days when

---

13 For this list of the hand's capabilities, see Raymond Tallis's *The Hand*, 22–8.

14 In "Letter to the Blind" from 1794, Denis Diderot writes: "The blind man has no other object, but by touch. He knows, from the account of others, that objects are known by means of the sight, as to him by touch. He farther knows, that there is no seeing one's own face, though it may be touched. He must therefore conclude sight to be a kind of touch …" (7).

15 The metaphor or image of the eyes in the peacock feathers attracted Michel Serres in his *Five Senses: A Philosophy of Mingled Bodies*. See the "Hermes and the Peacock" section of chapter 1, 38–57.

visitors could walk up to a hung painting and run their hands over its surfaces. Alarms go off and guards rush over. Constance Classen in her study of touch and museums helpfully reminds her readers that, in the early museums touching was "so commonplace as to customarily escape mention" (*The Deepest Sense* 137).[16] She gives many examples, among which is one noted in 1710 by Zacharias Conrad von Uffenbach, who recalls that before an attempted robbery of the crown jewels from the Tower of London, these treasures were not beyond the reach of the fingers, that it used to be possible "to get one's hand through the grating and pick up the articles to feel their weight" (*The Deepest Sense* 137). Like the wild animals once also on display in the Tower of London's menagerie, admirers could reach through the bars and touch what they enclosed. Classen also mentions the remarks of the South Kensington Museum's director, who, in 1860, complained: "We have great difficulty in preventing them [the multitude of visitors] expressing the emotions they feel in looking at a picture" by touching it and consequently scraping off little bits of pigment (*The Deepest Sense* 145).[17] (Now, I suppose, this impulse is honored through remote means, such as pressing a button placed next to the exhibit, which activates a digitally driven informational display. This device has become a substitute for touch.)

The absence of direct touch in the act of regarding an exhibited painting, though, does not mean that the tactile experience is unavailable, for viewing can have the characteristic of a "hands-on" experience. As several commentators on this phenomenon have remarked, the visual and the tactile sensations are closely wed so that even when we are looking at a painting, it is not clear that we are attending solely to its visual qualities. As one commentator puts it, the eyes rather than the hand "can function like the organ of touch" (Ahmed and Stacey 6). Perhaps aware of the tradition of the primacy of touch as well as taking their cue from Michel Serres's assertion that the senses are not islands that keep to themselves but are entwined with each other, theorists often remark on the ways in which the eyes function like the organs of touch. They understand that the haptic and the optical are not necessarily two sides of a dichotomy.[18] The two senses can "slide into one

---

16 See also Constance Classen's earlier essay "Touch in the Museum," which is part of her edited collection entitled *The Book of Touch*, 275–85. At The Victorian Tactile Imagination conference held at Birkbeck College, July 10–20, 2013, Classen delivered a paper, "Victorian Intimacies: Love, Death and the British Museum," which spoke of how nineteenth-century visitors to the British Museum and other collection sites satisfied their tactile cravings by stroking the antiquities, fingering the paintings, and probing the mysterious swathed bodies of mummies.

17 For this example, Classen has drawn from Richard Altick's *The Shows of London* (1978).

18 For instance, see the introduction to Michel Serres's *Five Senses: A Philosophy of Mingled Bodies* (7), and Claude Bouillon's *Skin: A Living Envelope* (6). More recently critics such as Hilary Fraser of Birkbeck College, University of London, have spoken about theories of the tactile imagination, and in particular, have recalled Bernard Berenson's *The*

another" (Marks xii). Steven Connor's remark that painting is "an amalgam … of seeing and touching" is apt (28).

Aware that painting actually aspires to the condition of touch, a painter consciously raises the tactile sense so that viewers experience touch through their eyes. The artist is conscious of a haptic visuality. Christopher Perricone, sensitive to this phenomenon, suggests that "Perhaps the great artist is he or she who lures us back to the recognition of the fundamental sense of ourselves by appealing to our most fundamental sense [touch]" (235). To support their observations, both he and Mark Patterson, in their fine reflections on the sense of touch, recall a passage from the writings of the late American art historian, Bernard Berenson (1865–1959), who, in his study of the Florentine painters of the Renaissance, suggested that the artist's task is to raise the tactile sense and embrace the viewer in it. For a painting to be effective, the painter must tap the tactile imagination. The varying muscular sensation inside the viewer's palm and fingers should correspond to the projections of the represented figure. Berenson observed:

> Now painting is an art which aims at giving an abiding impression of artistic reality with only two dimensions. The painter must, therefore, do consciously what we all do unconsciously—construct a third dimension. And he can accomplish his task only as we accomplish ours, by giving the tactile sense, for I must have the illusion of being able to touch a figure. I must have the illusion of varying projections of this figure, before I shall take it for granted and read, and let it affect me lastingly. (As quoted in Patterson 86)

There are, of course, many ways in which the artist constructs this "third dimension" (this sense of touch), and lures the viewer into the tactile realm. One way is through the actual representation of the hand, which serves as the sensation's envoy. I have often wondered whether the emphasis upon the depiction and placement of hands in portraits is not partially because they represent the sensation and evidence of touch. It is interesting that John Burnet in his *Practical Hints on Portrait Painting* (1850) emphasizes the importance of hands in portraiture and proclaims, "The treatment of the hands in a portrait shows the invention of the painter perhaps more than any other part of the human figure" (37). For him, as well as for many other artists who were aware of its importance in relaying a person's texture and presence in the world, the painted hand signifies a haptic knowledge of the subject's body as well as what surrounds and interacts with it. The hand's presence in the work of art triggers the tactile experience in the viewer's memory, promotes a sense of direct contact, and replicates the intimacy associated with it. Victorian novelists and poets knew this device well. It was all part of what has recently been referred to as the Victorian tactile imagination. Adding to Helena Michie's discussion of the lovers' hands in *The Flesh Made Word*, recent

---

Florentine Painters of the Renaissance (1896), which proposes that every time our eyes recognize reality, we are, as a matter of fact, giving tactile values to retinal impressions.

work on the novel by such critics as William A. Cohen, Peter J. Capuano, and Bruce Robbins have addressed the metonymic function of hands in Victorian literature and the meaning of such sensory encounters.[19] And one cannot forget Alfred Tennyson's *In Memoriam A.H.H.* and the frequent attention to the touch of the hand as a metonym for both Arthur Henry Hallam's presence and absence.[20]

Yet another means of creating a tactile response to a painting is by showing the way in which the subject is touched by light and shadow. These illuminations and shadings allow for the texture of touch. Shadows cast from interrupted light give a tactile dimension to a two-dimensional work. They are evidence of some physical presence that lurks near the subject and draws close. Shadows caress the flesh and seduce the viewer by defining the body's shape and allowing the eyes to touch its fleshly volume.[21]

Undoubtedly, however, one of the most significant ways in which a painting aspires toward the condition of touch and proffers a tactile experience is through its very medium. A canvas is itself a surrogate for skin, and the skin itself is similar to the canvas, for it too can be painted and function as a surface waiting to receive the artist's brush. For instance, there are parallels between painting a canvas and coloring the face (cosmetic practices).[22] Applying makeup/color and shadow on the skin is not that dissimilar from the artist who strokes the canvas with a brush and adds colors to its surface. Both leave a sign of the hand. Both are evidence of the manual touch. Traces of the fingers' impressing force mark and chart their

---

19 See, for instance, Peter J. Capuano's "Handling the Perceptual Politics of Identity in *Great Expectations*" which discusses Dickens's obsession with hands in that novel; William A. Cohen's *Embodied: Victorian Literature and the Senses* (2009); and Bruce Robbins's introduction, "The Secret Pressure of a Working Hand," to *The Servant's Hand: English Fiction from Below.* At the NAVSA 2013 Conference in Pasadena, California, there were several panels remarking not only on the phenomenon of touch but also on the hand's role in various Victorian texts. One panel, assembled by Kimberly Cox of the State University of New York, Stony Brook, featured papers on "Novel Hand Plotting," "Handling Desire," and "Handling and Handwriting: Late Nineteenth-Century British Art Critics as Graphologists."

20 An example can be found in Lyric X when Tennyson poignantly reflects on the absent hand and worries that the boat carrying Hallam's body back to England might wreck. The speaker fears that the "… hands so often clasped in mine, / Should toss with tangle and with shells" (ll. 19–20).

21 For an excellent book on the concept of the shadow, see Victor L. Stoichita's *A Short History of the Shadow*. Steven Connor's remark in *Thinking Through Skin* is worth noting here: "The flesh displayed in posters and magazines … looks touchable, caressable, for what is to be impressed upon our eyes is the way it has been touched by light" (as quoted in Ahmed and Stacey 38).

22 In her discussion of the nineteenth-century realistic novel within the chapter "Mirror of the Soul: The Epidermis as Canvas," Claudia Benthien remarks that these novels "employed the classical color code of painting" and produced a "literary painting" (103). She discusses the literary technique of "creating a semiotics of character types by means of the skin's structure and shadings" (104).

creation. As Serres observes: "The painter, with the tips of his fingers, caresses or attacks the canvas" (35). The brushstroke is evidence of the painter's manual touch, a sign of the maneuvering hand, which brings the subject into a visible form. In this respect, Patterson beautifully suggests that the artist "touches, he feels, he reckons weight, he measures space, he molds the fluidity of atmosphere to prefigure form in it, he caresses the skin of all things" (87). With the language and pressure of touch the artist composes the language of sight. The skin becomes visible. Through his eyes the viewer figuratively fingers the skin/canvas. Here one is reminded of Serres's sensitivity to the tactile art of Pierre Bonnard's studies of nudes. When Serres gazes at one of these paintings, he understands that Bonnard is not so much appealing to sight as to touch. His portraits capture "the feeling beneath the fingers" (30) and allow the viewer to "touch the skin of things" (35).

Two Victorian Paintings

The context of this compelling haptic visuality, coupled with the impulse to touch wild skins (discussed in the chapter's opening), creates a significant opportunity to look carefully at two Victorian paintings of human figures posed on animal skins: Sir Lawrence Alma-Tadema's *In the Tepidarium* (1881) and, perhaps surprisingly for those familiar with the work, James McNeill Whistler's *Symphony in White, No. 1: The White Girl* (1862). For the purposes of convenience and to conform to accepted practice, I shall refer to the latter work as simply *The White Girl*.[23]

Both paintings portray an intimate moment of contact between a wild animal's and a woman's skin; each, in a sense, taps into the desire to finger the enclosed, exotic animal's hide, as discussed earlier. And each participates in the cultural anxiety surrounding the confusion over the more fluid boundaries between human and beast following the popularization of evolutionary theory and the subsequent dissolution of a fixed and immutable division between them: the hand was once a paw; human skin had once been animal skin. The eroticism of these paintings partially depends upon the anxiety emanating from Darwin's sense that we (humans and animals) may all be melted together. Touch between animal and human is fraught as well as pleasurable.

Gazing at these paintings selected from Alma-Tadema's and Whistler's *oeuvre*, one wonders exactly how these works of art replicate the sensation of touch. To what extent do these paintings permit the observer to dissolve his subjectivity into the subject? What have these two artists done to prompt viewers to reach into their works with their eyes so that they can virtually extend their hands through the frame and feel the animal skin not only caressing the subject's body but their own as well?

---

[23] The hyphenating of Alma Tadema is the more common practice, even though there are many instances of the artist's name not being hyphenated.

Presented through the erotic lens, both Alma-Tadema's and Whistler's portraits express the desire to place a hand upon the wild other, to have direct access to the feel or the sensation of it. Both pieces participate in a skin-to-skin fantasy, which, of course, is still very much alive, and continues in the popular imagination to relegate touch to the lowest, basest position by equating it with the more carnal forms of love. Recently, for instance, I happened upon a popularized adaptation of this fantasy when I stepped into a men's room of a restored hotel in downtown Buffalo—the Lafayette. (I had wandered in to see how it was designed.) I was intrigued to find hanging on the wall a 1950s oversized poster of a scantily clothed female astride a polar bear skin. Her extended and searching fingers "caress" the polar bear's head (even though he is skinned, his face registers the delight of her touch). The picture invites the viewer to feel what it must be like to have her naked skin intimately rubbing against the animal's fur (and perhaps the observer's).

Figure 4.8 "The Polar Bear Rider"
Source: Lafayette Hotel, Buffalo, NY.

Although, in their own way, the two Victorian paintings I am going to discuss are earlier versions of this cruder image, I am sure that neither Alma-Tadema nor Whistler would want their pieces placed alongside such an explicit and trite, as well as badly executed, image. I apologize.

In the Tepidarium

Alma-Tadema's *In the Tepidarium* is a politely erotic painting showing a fully naked woman lying, completely in view, on a bear's skin draped over a slab of classically veined marble within a chamber of a Roman bath. (A tepidarium is the room situated between the hot and cold chambers.) The model's figure listlessly lies, exposed, upon the bear's lush fur.[24] As if extending the subject's languid sexuality, the fur spills over and touches the floor. Both model and hide recline upon a slab of marble, which, like human skin, exhibits the veins running through it. Each of these surfaces touches the other so as to set up a sort of *frisson*, encouraging the viewer not only to feel (through the eyes) the textures of the painted scene but also to experience the intended attendant tactile pleasures of the skins' intimate contact. Each skin, whether bodily or architectural in nature, brushes up against the other. Through the skill of his brush strokes and his own touching hand, Alma-Tadema *attempts* to seduce the viewer by letting his model's naked body become a site/sight for touching. With an intricate set of pigments required to depict the flesh colors, he has painted the skin of the canvas in order minutely to replicate the skin of the model, the bear's hide, as well as the architectural skin of the marble slab.[25] Through his skill, Alma-Tadema encourages his viewers not only to survey but also vicariously to touch his model's exposed body and visually run their fingers over it surfaces. Indeed, Alma-Tadema often invited visitors to his studio to look at the painted skins through a magnifying glass. The visible contrast among the long, rough, irregular fur of the wild bear, the smooth flawless skin of the woman, and the evenly polished marble contribute to what tactile experience there is. And so do the attendant sensations of warmth and chill: the heat of the living body; the warmth of the fur; and the coldness of the marble.[26] The feel/temperature reflects

---

24 The bear rug was a prop taken from Alma-Tadema's studio. In photographs of his studio, one can catch sight of an assortment of bear and leopard skins either on the floor or draped over the backs of chairs.

25 Based upon reading books about portraiture, I understand that skin poses the greatest challenge to artists. The number of colors required accurately to represent the subtle gradations of flesh have resulted in a plethora of treatises on skin color. Among these are Henry Murray's *The Art of Portrait Painting* (1851); and *Progressive Lessons Intended to Elucidate the Art of Portrait Painting in which is introduced a systematic arrangement of the Colours and Tints used in Flesh, Draperies, and Back-Grounds with Examples* (1824).

26 Russell Ash, in his study of Alma-Tadema's paintings, notes that though Alma-Tadema was able to convey almost any texture, from fur to feathers, it was his painting of marble that singles him out. Helen Zimmern's study of Alma-Tadema also mentions his excessive scrupulousness. When, for instance, he painted a tiger skin for one of his

the tepidarium itself, a space, as I already explained, that lies between the hot and cold rooms of a Roman bath. Alma-Tadema adds to the potential eroticism of these sensations by placing the image of the phallic strigil (an instrument used for scraping the skin after soaping and oiling it) in her eager hand and by meticulously rendering the soft ostrich-feather fan the model teasingly holds to cover, yet expose, her pubic hair.[27]

However, for all the painting's teasing references to touch, and particularly to the intimate proximity of animal and naked human skins, *In the Tepidarium* is not ultimately erotic: it is only what critics recognize as "politely" so. It is refined. The artist's desire to make this touching the subject of his painting is not fully realized or is, as some might say, potted and impotent. To begin with, the artist condemns his viewers to a perspective, which keeps them at a distance. Spectators are obliged to survey the model's entire figure and, consequently, remain more or less aloof. They are as detached as is the model, who has no eye for the observer. This distance hinders any strong subjective response from the viewer so that rather than move in to the touch, his eyes glide over the various skins' exteriors and slide along the canvas's smooth surfaces. There is nothing to interrupt this surveillance, for these surfaces are too perfect. In particular, the model's skin displays no mark, scar, crack, or chip. It is invulnerable and, perhaps more to the point, impenetrable. Fetishized into an object, which is as impervious as the cold veined, solid marble upon which she rests, her skin cannot be wounded. It promises no depth, fold, or hint of weakness. It does not portray a permeable living envelope. Only the draped flayed bear skin with its lowered, defeated head seems to draw close to the idea of the wound, to the possibility or memory of penetration. Furthermore, because of these smooth surfaces, nothing interferes to trip the viewers' eyes so that they desire to feel beneath and in between the shadows or lift the ostrich feathers away. How different from Rubens's painting of his Heléné Fourment's naked flesh, which, through its unevenness and imperfections, invites the viewer to feel beneath and in between the shadows created by her lush bear wrap. Combined with her own roving eye, which seems to touch the observer, the tactile sensation of the animal skin caressing and partially concealing her nakedness is tantalizing. One senses the depth of touch, not the play of surface.

Significantly, when John Ruskin commented on Alma-Tadema's paintings, he suggested that the artist's "technical perfection was in inverse ratio to the impotence of the subject he was portraying" (as quoted in Ash n.p.). Ruskin was

paintings, *An Audience of Agrippa's*, he gleefully asked a viewer, "Don't you see him wag his tail?" (56).

27 When Pears Soap Company considered using this work to advertise its soap (the woman does have very clear and clean skin), they were worried about the painting's sexual suggestiveness and decided not to use it. According to Russell Ash, it was also rumored that Edward VII, after seeing this painting, commissioned Alma-Tadema to produce a series of more avowedly pornographic works (*Sir Lawrence Alma-Tadema*). These paintings are said to be hidden behind a curtain in Windsor Castle.

Plate 3 Sir Lawrence Alma-Tadema's *In the Tepidarium* (1881)
Source: Lady Lever Art Gallery. National Museums Liverpool.

not alone in faulting Alma-Tadema's paintings for this weakness. Both Edmund and Ellen Gosse criticized him for fetishizing surfaces and, thereby, emptying the subject of its life. In her recent book, *Lawrence Alma Tadema* (1990), Louise Lippincott recalls one of Ellen Gosse's critical remarks: "There can be no doubt that Mr. Alma Tadema is so fascinated by the beauty of details and of surface-painting, and so accomplished in rendering it, that he sometimes loses sight of the original intention of his picture [the eroticism]" (73). Similarly, Edmund Gosse, in his 1882 article on Alma-Tadema for *Illustrated Biographies of Modern Artists*, also commented on the artist's narrow and technical attention to the "brilliant reproduction of surfaces" (76). Alma-Tadema is too involved with the visual details. He concentrates too tirelessly on the subject's surfaces. The consequence is that the visual and the ornamental details of his paintings run the risk of overwhelming the spectator's eye and compromising or interfering with what tactile sensation there could be. These criticisms seem to be responding to the lack of what William Hazlitt identified as "*Gusto*" in painting. They are recognizing that for all the attempts to appeal to the sensuality of touch, little in Alma-Tadema's work leaves an impression on a sense, as Hazlitt explained, so that it "excites by affinity those of another" ("On *Gusto*" 202). So trapped in their own visual façade, the paintings' details become isolated (perhaps self-referential) and fail to arouse or associate with the tactile experience. The woman has the look of flesh, but not the feeling of it. Alma-Tadema's reclining naked subject has none of Titian's flesh-colors that according to Hazlitt make Titian's women "sensitive and alive all over" ("On *Gusto*" 201). Furthermore, the marble also fails to reveal a tactile life within itself. When Hazlitt talked about the Elgin Marbles, he spoke of his admiration for their outer "skin" ("On the Elgin Marbles" 329) which reveals a life within it: "The skin (or the outside) seems to be protruded or tightened by the natural action of a muscle beneath it" (337). The consequence is that the viewer gazes at the image but does not experience what Hazlitt identified in "On *Gusto*" as a "tingling" (202). No "sting [is left] behind in the mind of the spectator" (202). Though *In the Tepidarium* might at first arouse a kind of *frisson*, the painting ultimately suffers from this absence.

Symphony in White, No. 1: The White Girl (1862)[28]

In contrast to Alma-Tadema's *In the Tepidarium*, the seemingly puritanical figure in Whistler's *The White Girl* is, strange to say, more potently erotic and tactile. Upon first glance this assertion might seem ridiculous, for there is none of the carelessly exposed full nakedness of Alma-Tadema's model. Instead, Whistler's mistress, Joanna Hiffernan, stands fully clothed, almost at attention, fully inhabiting

[28] *The White Girl* was initially rejected at both the Royal Academy and the Salon in Paris. Eventually, however, the painting caused quite a sensation when it was on display in the *Salon de Refusés* in 1863.

an 8-foot-long slender frame (84.5 in. x 42.5 in.).[29] She wears a lengthy white cambric dress, which extends from her neckline to below her feet; her clothing envelops her as if it were a loose second skin, leaving only a slight glimpse of her neck and hands. The heavy and extensive white curtains in the background suggest yet another covering, and contribute to the sense of her body being draped and hidden. As Whistler once impatiently remarked, "My painting simply represents a girl dressed in white standing in front of a white curtain" (Spencer 303). There is no opportunity, as there is in Alma-Tadema's work, for surveillance of the naked body. What little skin Whistler exposes is either neutralized or overwhelmed by the portrait's blanched tones. The subtle shading and fleshliness one finds in the skins of a Titian or a Rubens are absent. Furthermore, Whistler has chosen to blend one of her hands into the whiteness of the cloth and to arrange the other so that its fingers disappear into the folds of her dress. Carelessly limp, these fingers drop the white flowers.[30] No hand visibly clutches an erotic object as it does in Alma-Tadema's painting. Yet, for all the painting's lack of fleshliness and its attention to covering rather than exposing its subject, an active sense of touch and eroticism attends the painting. Whistler has honored the mandate that the painter's first business is to rouse the tactile sense; indeed, it is, paradoxically, the subject of this painting.

The question is: if there is no nakedness and no apparent skin-to-skin contact, and if there is not even an engaged hand, where, one wonders, does the tactile sense and eroticism reside? How does the viewer feel Whistler's and his subject's touch? One source is to be found in the fact that Whistler has devoted a large portion of the canvas's space to the exquisitely rendered white material of the dress and the drapery so that the texture and shadows of the cloth are alive; the eye fingers these and feels the folds, the wrinkles, the tucks, and the gatherings.[31] Obviously, however, the more sensual tactile experiences of the painting do not

---

[29] Whistler met Joanna Hiffernan (b. 1843) in 1860 while she was at his studio in Rathbone Place. During her six-year liaison with him, she modeled for some of Whistler's best-known pieces. She was in France with him during the summer of 1861, and in Paris during the winter of 1861–1862 at which time, at a studio in Boulevard des Batignolles, she sat for *The White Girl*. Although Whistler had other mistresses, Hiffernan continued to take care of him as late as 1880.

[30] Apparently Whistler instructed Hiffernan to hang her arms listlessly and maintain an expressionless face in order to exclude the idea of narrative.

[31] In an instructive letter (November 1881) to Thomas Delano, Whistler speaks of the importance of shadow. He writes: "Continue now with your flesh color—painting *from the light* into the shadow while the shadow is wet—so that you will really be covering the whole head in one sitting and indeed with one painting—This Tom is the simplest and best way out of it all—you will mix your flesh tints of course with white—and as you get towards the shadow you will see how much darker the grey or brown looks than nature, and then you will perceive the color that there is in shadow and you will be enabled to reach that by a mixture of your grey with some of the flesh tone on your palette—and so my dear Tom you will proceed and finish" (*The Correspondence of James McNeill Whistler*).

Plate 4 James McNeill Whistler's *Symphony in White, No. 1: The White Girl* (1862)

Source: National Gallery of Art, Washington, DC.

emanate from this sweep of whiteness. Rather, they originate from the contrast between that diffused paleness and the mistress's loosened red, lush hair lying over her shoulders, and, even more vibrantly, from the jarring dissimilarity between the smooth "civilized" woven cloth and the darker wild hairy texture of the wolf's fully visible face (indeed, a face with features, which suggestively mirror the model's own). But most provocatively, the excitement and exoticism reside in what is suggested rather than seen—from the *possibility* that his model is bare footed (she has after all let her hair down) and with her toes is caressing and feeling the wolf's wild fur. (Whistler actually foregrounds the wolf so it, more directly than his mistress, reaches out to touch the viewer.) Hazlitt's "tingling," comes from this probable skin-to-skin contact. The *Gusto* of this painting resides in this imagined encounter, secreted and secluded beneath the long skirts of the "White Girl's" dress. There is, of course, no way to lift up her skirt and see. The possibility of this hidden tactile moment is stronger and more sensual than what is directly on view. And so too is the eroticism of the mistress's body and skin concealed within cloth and neutralized by the surfeit of white coloring.

Why, then, is this painting so much more erotic, dangerous, and tactile than the more fully exposed *In the Tepidarium*? Why is the sense of touch strong and stimulating, even when it is in hiding? I am reminded of a passage from Tallis's *Michelangelo's Finger* (2010), in which he suggests that the experience of touch "goes beyond what is currently appearing to us" (122). As he puts it, touch is not limited to direct tactile contact; rather, it also resides in the possibilities inherent in that contact. Through the specter of touch a person is "led to believe in the existence of something that is other than the sensation that [he or she is] feeling" (120). A person is "in touch with more than [he or she is] currently touching" (122). Perhaps I am taking too many liberties with Tallis's remarks, which are prompted by his sensitivity to the mysteries of cognition, but they do suggest why, in Whistler's painting, the viewer, whose sense of touch has been aroused by the textures of the painting's drapery as well as the wolf's hide, is stimulated to feel the possibility of the strength and pleasure of that hidden tactile caress between the model and the wolf. The viewer is all too aware of the sensations emanating from this contact. Tallis's observation also helps explain why, when the viewer regards the one exposed hand of his mistress, he or she is invited to sense the potential sensuality of that hand. *The White Girl* seems to draw its power from these possibilities. Touch in this painting is not as much direct as it is a possibility.

In a sense, Whistler's painting realizes the yearning of those zoo visitors who wished to have their fingers licked by wolves; it makes tangible a desire to feel their own wildness and attach themselves to an exotic other, especially the wolf, known for its rapacious and aggressive qualities. The hidden pleasure of this contact fills the painting with anxiety, and, as such, adds to its eroticism. This suggestive touching challenges the supposedly definitive boundaries between humans and animals; it participates in the cultural anxiety surrounding the confusion over the more fluid boundaries between human and beast. Rather than having shed the animal skin, the "White Girl" in Whistler's painting is not complete without it. The wolf's skin supports her and is part of her. She is a hybrid; she has not separated from the beast. As if parodying Botticelli's *Venus*, she rises from and is still attached to this wildness. Her whiteness and her long dress, her second skin, hide the fur underneath. She digs her toes into the wolf and faces the viewer, who stares nervously, yet pleasurably, at a representation of what those used to believing in the ultimate division between animal and human would rather not recognize. Art critics who see Whistler's painting as an "allegory of innocence" or as an "allusion to the Virgin Mary," would rather not acknowledge that her folded hand might also be a paw, and that her features and the wolf's are similar.[32]

This observation takes me back to Alma-Tadema's *In the Tepidarium*. For all its alluring nakedness, this painting is not fraught with this cultural and sexual anxiety, for the nude female figure and the bear rug are not joined. Instead, they are layered. There is less of a sense that the bear rug is a part of her, that she digs her fingers or her toes into it. Indeed, in a way, she has rid herself of this history so that she can expose her untainted human flesh. (Why, otherwise, would Pear's Soap Company, in the late nineteenth century, have considered using this painting as part of its advertising campaign?) The bear skin lies beneath primarily as a decoration. (Alma-Tadema put this bear skin in at the last minute to fill up an empty space in the composition.) The bear is flayed and defeated beneath her: note that its bowed head does not stare out to entice the viewer, as does the wolf's, but rather sinks, acknowledging defeat. The bear's skin, then, is more of a prop or a cliché of ornamentation; it is not the touchstone of her existence.

Conclusion

The tactile experience is in varying degrees a factor in both paintings, but in Whistler's the possibility of touch stimulates a more powerful and anxious response. This difference between the two pieces suggests something paradoxical about the nature of touch. It implies that the sensation of touch is not necessarily located in the actual tactile encounter. Rather, at times, the experience of touch occurs before the fingers caress or impress another skin: it resides, instead, in the action of extending the naked hand toward something other than itself. Rather than depending on the moment of actual contact or encounter, the verve and *frisson* associated with touch reside in the intermediary space of possibility or fantasy, which teasingly and mysteriously linger between the reaching hand and

---

32 Beryl Schlossman, in her *Objects of Desire: The Madonnas of Modernism*, is one critic who sees the painting as an allegory of lost innocence. Critics who wish to spiritualize *The White Girl* often reflect John Burnet's fears concerning the rendition of hands when painting a portrait (referred to earlier in this chapter). As were so many others in mid-nineteenth-century Britain, Burnet was anxious about the possible links between beasts and people and therefore wanted to make sure that the artist paid attention to the distinctions between a human's hands and a beast's paws (*Practical Hints on Portrait Painting* 35).

its object. As Connor remarks, "The hand is the body's possibility" (140). And as Tallis observes, "To touch is to awaken to awareness" of something beyond the self (*The Hand* 31). The sensation lies in the anticipation of that contact. With this consideration in mind, one can turn one's attention back to those visiting nineteenth-century menageries and zoos (as well as those viewing the "real Indians" in the assembly halls of Manchester), who were eager to push their hands through the figurative or real iron bars so as to touch the enclosed exotic creatures. Consider the anticipatory moment depicted in Figure 4.1 in which the zoo visitor's extended hand reaches out toward the rhinoceros. The excitement, pleasure (that "tingling" Hazlitt describes), and daring associated with this gesture, as well as with others, was, perhaps, not so much indebted finally and actually to stroking the wild animal's fur or skin as they were beholden to the feelings suspended in the space between the extended fingers and the sensation of the other's skin. Strange to say, the experience of touch resides in the prospect of it (whether the caress is realized or not). Touch is elusive; it is both physical and subjective. As our deepest sense, it lies everywhere.

Chapter 5
Wild Skins and Mapping the Victorian Landscape

So Geographers in Afric-Maps
With Savage-Pictures fill their Gaps;
And o'er unhabitable Downs
Place Elephants for want of Towns

Jonathan Swift[1]

A Preface

Because this study of wild animal skins has intermittently addressed Victorian Britain's preoccupation with its displays of animal hides and mounted trophies, this book fittingly concludes by considering how this fixation infiltrated the ways in which Victorians mapped the world about them. In order to explore this aspect of the Victorian cartographic imagination, I have divided the chapter into three sections.

Part One, "*Tess of the D'Urbervilles*," introduces the connection, both real and metaphoric, between these wild skins and the mapping of the Victorian landscape. It demonstrates how Thomas Hardy summoned images of skin in order to trace the contours of Tess's tragedy and chart the topography of her downfall. With his many references to the complexion of the earth, to the heroine's wounded skin, as well as to Tess's feral nature (her skin is a trophy), Hardy, Victorian to the core, partially plots the course of this novel by employing the idiom of the type of colonial map that uses the iconography of the hunting and collecting of wild skins.

Part Two, "Animals and Nineteenth-Century Maps," steps back from Hardy's particular use of this idiom in order more generally to examine the Victorian adventurers' maps that take their cue from hunters' encounters with animals. In addition, this section also discusses the period's fascination with zoogeography, a cartographic genre popular in the mid-century among scientists desiring to chart the distribution of animals throughout the globe. As products of a colonial culture intent on ordering alien territories, the maps drawn by scientists and adventurers seek to assert control by transforming the messiness of empire (discussed in Chapter 2) into systematic, two-dimensional cartographic documents. In particular, the more scientific zoogeography presents a solid and static schematized view of the world in which animals stand alone (as if behind bars at the Belle Vue Zoo), and remain in their "proper" places, isolated from humans.

---

1 From "On Poetry: A Rhapsody" (1733), lines 177–80.

These cartographic styles, however, are significantly different from the more sensuous geography that orients Gerard Manley Hopkins's inclusion of animals and their skins when mapping/describing his surroundings. Part Three, "Gerard Manley Hopkins: Animals in the Landscape," is about this contrast. This section dwells upon passages in Hopkins's diaries, notebooks, and journals, which appear to transform the maps representing classified zoogeographic spaces into living, breathing, and three-dimensional experiences. Although Hopkins was occasionally indebted to the conventions associated with formal cartographic practices, he also chose to break away from their models so that he could, in a manner even more intricate and intense than Hardy's, not only acknowledge the body in the landscape, but also honor the defining and navigating sensations of motion and touch. Hopkins's use of wild animal skins reveals a different kind of map—as unconventional as his famously idiosyncratic syntax. His descriptions of landscape proffer an animistic reading of nature and the world. No longer is the globe seen in terms of a captured trophy or an isolated species in a particular region; rather, his surroundings are mapped through motion, touch, and a vibrant intermingling of the animal and the human.

Part One: *Tess of the D'Urbervilles*

Before more generally discussing the use of animals and skins in mapping the Victorian landscape as well as Hopkins's transformation of this cultural practice, it is beneficial first to be reminded of this cartographic perspective in Thomas Hardy's 1891 *Tess of the D'Urbervilles*.

The Cartographic Perspective in Tess

Recalling the fictional map which prefaced the Wessex Novels edition of 1895–1896, Hardy opens *Tess* with a conventional map-like description of the novel's setting. He traces the landscape from the view of one walking its land and occasionally elevating his eye so as to give a bird's-eye perspective of the surroundings. Moreover, within that description's closing phrases, as if evoking the ancient practice of drawing maps upon parchment (the treated hide of an animal), Hardy also summons the image of the landscape's skin in order to plot the contours of Tess's physical being. The "dark green threads" overspread the epidermis of the landscape in much the same way Tess's blue veins lace their way through the "beautiful feminine tissue" of her naked arms and neck (56).[2] Hardy's description of the village of Marlott maps the earth's skin and simultaneously anticipates the delicately veined skin of Tess, who is the daughter of this land.

2 When describing Tess, Hardy draws attention to "the blue veins of her temple" (169).

> The fertile and sheltered tract of country, in which the fields are never brown and the springs never dry, is bounded on the south by the bold chalk ridge that embraces the prominences of Hambledon Hill, Bulbarrow, Nettlecombe-Tout, Dobury, High-Story, and Bubb Down. The traveller from the coast, who, after plodding northward for a score of miles of calcareous downs and corn-lands, suddenly reaches the verge of one of these escarpments, is surprised and delighted to behold, extended *like a map* beneath him, a country differing absolutely from that which he has passed through. Behind him the hills are open, the sun blazes down upon fields so large as to give an unenclosed character to the landscape, the lanes are white, the hedges low and plashed, the atmosphere colourless. Here, in the valley, the world seems to be constructed upon a smaller and more delicate scale; the fields are mere paddocks, so reduced that from this height their hedgerows appear a network of dark green threads overspreading the paler green of the grass. (12 [italics mine])

Continuing to call on the cartographic image, Hardy also draws the reader's attention to the land's traditional boundaries, especially when speaking of the women's "walking club," which features women participating in the ritual of marking the bounds.[3] Within these confines wanders the vulnerable Tess. Though the opening passage paints a softer picture of the landscape, the novel shifts to reveal a far harsher environment. Tess is soon enveloped within a landscape mapped by images of wounded and coarse skin. For example, when Hardy describes the Flintcomb-Ash swede-fields, in "Phase the Fifth," he focuses on the "complexion" of the desolate, dark landscape where Tess is condemned to labor slavishly, night and day:

> The swede-field in which she and her companion were set hacking was a stretch of a hundred-odd acres, in one patch, on the highest ground of the farm, rising above stony lanchets or lynchets—the outcrop of siliceous veins in the chalk formation, composed of myriads of loose white flints in bulbous, cusped, and phallic shapes. The upper half of each swede-turnip had been eaten-off by the live-stock, and it was the business of the two women to grab out the lower or

---

[3] Hardy comments on this ritual:

> The forests have departed, but some old customs of their shades remain. Many, however, linger only in a metamorphosed or disguised form. The May-Day dance, for instance, was to be discerned on the afternoon under notice; in the guise of the club revel, or "club-walking," as it was there called.
>
> It was an interesting event to the younger inhabitants of Marlott, though the real interest was not observed by the participators in the ceremony. Its singularity lay less in the retention of a custom of walking in procession and dancing on each anniversary than in the members being solely women … The club of Marlott alone lived to uphold the local Cerealia. It had walked for hundreds of years, and it walked still. (13)

> earthy half of the root with a hooked fork called a hacker, that it might be eaten also. Every leaf of the vegetable having already been consumed, the whole field was in colour a desolate drab; it was *a complexion without features*, as if a face, from chin to brow, should be only an expanse of skin. The sky wore, in another colour, the same likeness; a white vacuity of countenance with the lineaments gone. So these two upper and nether visages confronted each other all day long, the white face looking down on the brown face, and the brown face looking up at the white face, without anything standing between them but the two girls crawling over the surface of the former like flies. (285 [italics mine])

Caught between the drab facial tones of the sky and the dreary skin colors of the expansive fields below, Tess is subject to their coarse features. Her own skin is sacrificed to the stubble that cuts and scratches her naked flesh. Glass splinters of the wind injure her exposed arms and shoulders. Attempting to protect her soft hands, Tess dons a pair of buff leather gloves (another layer of skin) but to minimal effect:

> Then stooping low she moves forward, gathering the corn with both hands against her knees, and pushing her left gloved hand under the bundle to meet the right on the other side, holding the corn in an embrace like that of a lover. She brings the ends of the bond together, and kneels on the sheaf while she ties it, beating back her skirts now and then when lifted by the breeze. A bit of her naked arm is visible between the buff leather of the gauntlet and the sleeve of her gown; and as the day wears on its feminine smoothness becomes sacrificed by the stubble, and bleeds. (88)

Like a cartographic diagram, the novel unfolds the routes of Tess's existence and traces the ways in which her skin is touched and ravaged by this hostile world. Indeed, the shape and progress of the novel call to mind nineteenth-century physical maps, which display the earth's surface, as if it were the epidermis of the world, by revealing its scales, layers, hollows, wrinkles, furrows, and wounds.[4]

More to the point, however, *Tess*'s map takes its cue from colonial adventurers who, as I shall discuss, chartered their way through bewildering exotic spaces by marking where they had captured wild skins. Within the territory of Hardy's fiction, Tess becomes one of these hunted feral beings. She is a "genuine daughter of Nature" (120), whose eyes stare at Alec D'Urberville "like those of a wild animal" (55), whose body brushes up against the fur of the leaves, and whose anxieties wait "like wolves" (195) to pounce upon her. Her hide is tracked and

[4] Michel Serres in *Five Senses: A Philosophy of Mingled Bodies* speaks of skin as being an inscribed map which, like the surface of the earth, makes visible the "wear and tear, scars from wounds, calluses, wrinkles and furrows of former hopes, blotches, pimples, eczema, psoriasis, birth-marks" (24). (Given the context of this study of wild skins, one notices, in Figure 5.1, that the earth's surface resembles the folds of a rhinoceros's hide.)

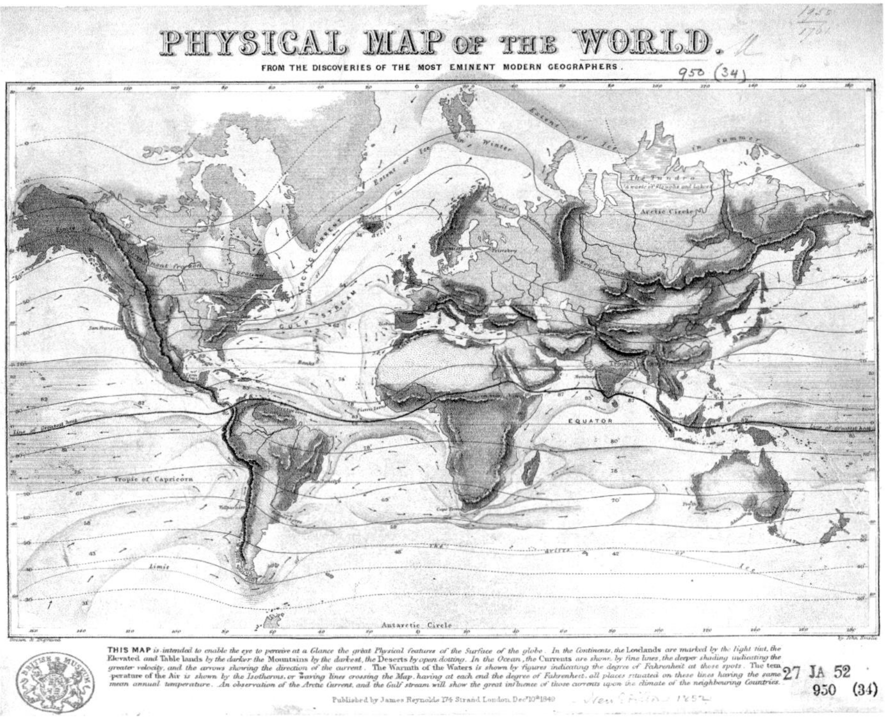

Figure 5.1 Physical Map of the World from the Discoveries of the Most Eminent Modern Geographers
Source: British Library Board 1849.

hunted. Like a snake, a leopard, and a wounded pheasant,[5] she suffers from those who desire to expose the delicate skin underneath her clothing, penetrate its alluring texture, and sacrifice or constrain her.[6] She is subject to an eye that stalks her so closely that it sees "the red-interior of her mouth as if it had been a snake's."[7] Even when she hides in a mound of leaves, her fate resembles that of the rabbits, hares, snakes, rats, and mice that retreat from the reaper and find their

5 For the wounded pheasants episode, see Hardy 278–9.

6 In an interpolated comment, Hardy ends "Phase the First" of *Tess of the D'Urbervilles* by wondering: "Why it was that upon this beautiful feminine tissue, sensitive as gossamer, and practically blank as snow as yet, there should have been traced such a coarse pattern as it was doomed to receive …" (74).

7 Hardy writes: "She had not heard him enter, and hardly realized his presence there. She was yawning, and he saw the red interior of her mouth as if it had been a snake's. She had stretched one arm so high above her coiled-up cable of hair that he could see its satin delicacy above the sunburn; her face was flushed with sleep, and her eyelids hung heavy over their pupils" (169).

refuge ephemeral (87)—she is discovered and destroyed. She is an embodiment of the white hart, which, according to legend, was eventually killed by hunters, in spite of Henry III's command that it be spared. Appropriately, the vale in which Tess's family lives takes its name from this tale.[8] Throughout, her enclosed body is subject to a chase; her skin is a trophy. She is a "hunted soul" (277).

Part Two: Animals and Nineteenth-Century Maps

With its spatial orientation centered on Tess as a trophy skin as well as on the sites of Tess's struggles—not to mention the novel's references to the landscape's complexion—the course of *Tess of the D'Urbervilles* reveals its indebtedness to maps drawn by hunters and explorers, who charted their course through their encounters with wild skins. *Tess* emerges from the shadows of these Victorian adventurers' maps in which trophies of animals functioned as landmarks.[9] In their cartographic sketches, explorers noted not only where they had slashed their way into an unfamiliar land but also where they had made incisions into a creature's skin and measured its hide. The red lines on these maps representing their routes across unknown, distant lands suggest the blooded course of their cutting. As a result, in addition to the rivers, mountains, and tribal groups of these explored areas, memories of exact spots where colonials and adventurers had captured wild skins literally oriented their way and helped organize the space of uncharted territories.

In order better to comprehend the importance that Victorians gave to animals and skins in the mapping of empire as well as to gain a better understanding of why this aspect of cartographic imagination was to cast such a lengthy shadow (so much so that it reaches Hardy's fiction), it is helpful to begin the discussion on animals and nineteenth-century maps by first turning to early nineteenth-century maps of Africa showing regions of uncharted vastness. These bald areas were enticing to those imperialists desiring to explore and fill in the empty and bewildering spaces.

Betts's 1856 *Improved Educational Map of Africa* and W. Keeble's *Africa: Showing the Latest Discoveries* (published by the Christian Knowledge Society in 1853), available in the British Library, are two examples of maps displaying these immense uncharted regions. The coastal areas and parts of North Africa are surveyed in detail, but the extensive center of the map is uncannily vacant. A gaping emptiness haunts the viewer's eye. In the latter map, speculative sentences

---

[8] See Hardy: Chapter II, n. 2, 403.

[9] These maps appear to follow the lead of ancient documents, which mapped bewildering spaces by substituting animals for names. Victorian adventurers' maps followed the tradition of the older maps, referred to in this chapter's opening quotation from Swift's "On Poetry: A Rhapsody." They continued to use animals such as "Elephants" to mark and metonymically represent uncharted territories. Exotic mammals and birds "fill[ed] the gaps" of these relatively unknown regions.

such as "Mountains of the Moon Supposed to extend Across Africa from the Mountains of Kong to Abyssinia" run through the interior's empty spaces, above and below the equator. It is interesting to look at these and realize that they are, later in the century, to be filled in by those like David Livingstone, who walked for years (from 1841 until his death in 1872) through uncharted ground. Livingstone frequently recorded his journeys not only by identifying the rivers he drifted down and by naming the peoples he met but also by narrating stories of the animals and birds he encountered and whose skins he regularly shipped to British scientific institutions.[10] As for many other colonials who traced their way through distant lands, these collected skins served as the raw material—the basic markers of place—of his maps. Early on in his African journeys, for example, Livingstone transported skins and bones of exotic beings to Richard Owen in order to facilitate the study of anatomy. Eager to lend a hand in the cause of science, Livingstone partially traced his way by mapping the skin of natural history and sending back sketched and hand-annotated maps of various districts. He also charted his way through his own skin, particularly when a lion jumped on him and ripped his shoulder. This wound was a landmark.[11]

This cartographic mode was especially relevant when, in 1849, Livingstone abandoned the role of traditional missionary and traveled for several months with William Cotton Oswell, a wealthy English hunter and explorer, who, like many a sportsman before and after him, almost exclusively, literally and figuratively, mapped the landscape through tales of hunting, danger, and cunning. For these individuals, the presence of huge herds of elephants, pools brimming with hippopotami, and lions roaming through forests in the dark of night constituted the landmarks and the map of Africa.[12] Oswell's orientation is similar to that of another great hunter and explorer, Roualeyn George Gordon Cumming, already mentioned in previous chapters, who charted the landscapes he passed through by noting what animals he had killed and how he had skinned them. For Cumming, especially, the skins he had collected defined the country's complexion and topography. (Cumming is said to have sent back 30 tons of trophies from Southern Africa.)[13] Indeed, as already noted in Chapter 2, Cumming was distraught when

---

10 Francis Galton in his "Introduction" to W. Edward Oswell's *William Cotton Oswell: Hunter and Explorer, The Story of His Life* (1900) writes: "Sixty years ago the interior of South Africa was a blank on our maps, the modern knowledge of its geography being based for the most part on numerous explorations made since that date" (xiii).

11 In a December 26, 1845 letter to James Ridson Bennett, Livingstone describes the attack, which marked his way through a region near Kuruman and Lattakoo, by mapping the wound on his body and drawing a diagram of his torn flesh. See http://www.livingstoneonline.ucl.ac.uk/view/transcript.php?id=LP115.

12 For examples of the importance of wild skins for Oswell when mapping the southern interior regions of Africa, see *William Cotton Oswell: Hunter and Explorer, the Story of his Life*.

13 James R. Ryan, "Hunting with the Camera: Photography, Wildlife and Colonialism in Africa" in *Animal Spaces, Beastly Places: New Geographies of Human-Animal Relations*

these important landmarks (these collected skins) were occasionally consumed either by insects or vultures, as well as destroyed by inept natives. These ruined trophies threatened his map's orientation.

A glance at other "hunting maps" of the time demonstrates a similar mode of charting the unknown. Relatively obscure territories of southeast China, for instance, were mapped according to shooting districts. An 1893 *Map of the Shooting Districts between Shanghai and Wuhu*, published by H.T. Wade and R.A. de Villard, presents this unmapped region by noting the swamps, hills, and marshes where the wild fowl abound. Because the diagram is surveying new territory, the tight, practiced symbolism of a more refined map is not to be found. Instead, descriptive words and names of the wildlife chart the space and replace the traditional signs of a familiar region. Phrases such as "Good Shooting," "Grassy Country," "Low Rolling Hills," "Reported Good Shooting Country," "Good Pheasant Hunting in Nov.," "Wild Boar Hills," and "First Rate Shooting," inscribe the map's surface. Other areas are not marked, for they are deemed to be of "No Importance." Some are left blank, except for noting: "Country to be explored by future sportsmen" or "Mountainous Country Unsurveyed."

Appropriately, the established taxidermy firm of Rowland Ward (discussed in Chapter 3), which dealt in skins and based its business on stuffing and mounting trophies, occasionally published narratives of collecting or hunting expeditions. In keeping with the travel genre, these books featured rough sketches of explorers' routes as well as foldout maps pasted to or pocketed inside the covers of each book. The spatial foundation of these maps was dependent, to a significant extent, on the knowledge of where certain animals or birds had been shot or skinned. For example, in chapter 2, "Big Game Shooting, 1887," from *Seventeen Trips through Somaliland and a Visit to Abyssinia* (published by Ward), Major H.G.C. Swayne, who surveyed the areas in 1886, 1891, and 1892, inscribes his journey by recalling where he had bagged the beautiful skin of a large African antelope (a koodoo), which he had admired and coveted.[14] He notes the places where, for instance, he had shot and skinned a panther by firelight (95), or where, by the light of torches, he had skinned a rhino (127). Each of these experiences creates the cartographic narrative. He even records the time his own flesh had been touched and wounded (as had Tess's) by the aggressive nature of his surroundings. In the manner of other nineteenth-century hunters, for instance, he describes, as had Livingstone, the moment when a lioness had punctured his flesh (Swayne 104).[15] At the end of his account, Swayne draws the reader's attention to the tracks in the dirt made

---

(Philo and Wilbert 203).

[14] Swayne writes: "Nür Osman and the Midgán led the way, carrying the head and skin of my first Koodoo, at which I could not help looking admiringly from time to time, for it was a great prize" (45). Just like the other skins he collects, the koodoo's hide becomes a landmark.

[15] The letters quoted in W. Edward Oswell's biography of his father *William Cotton Oswell: The Story of his Life* (referred to earlier in this chapter) often feature the father's

by the caravan carrying the hide of a six-foot lion (219). Skin orients his way and contributes to the delineation of his travels.

Similarly in Henry Zouch Darrah's 1898 narrative, *Sport in the Highlands of Kashmir* (also published by Ward), collected skins mark his way and aid him in charting his eight-month trip in Baltstan and Ladah. The hides he brings home, like a map, make visible what his readers could not normally see or experience. One of the narrative's inserted maps, "The Great Bend of the Indus," displays circles which indicate exactly where a certain animal was shot and skinned. As if to emphasize the importance of this cartographic tool, Darrah concludes his narrative about his hunting exploits with chapters that concentrate on the best skinning knives to carry or the preferred methods of curing and cleaning skins. Like Cumming, he is concerned that these spoils that recall a memory of a particular location/skin will disappear if it is not properly cured. Worried that the map of his travels will evaporate, he closes his book by advising:

> As a general rule it is, I think, advisable to have cured at Srinagar the skins which the sportsman wishes to retain. He cannot well keep them with him, as his camp is constantly moving, and they would not only add a good deal to the expense of travelling, but would run the risk of being injured.
>
> There, in the summer months, the heat is considerable, and it is damp heat, the worst possible for skins. Consequently if they are left for any time uncured they are likely to go bad. So I think it is wiser to get the curing done. They can then be taken home at leisure, and properly dealt with there. If, however, the sportsman is only doing a short shoot, he may be able to take his trophies into Srinagar himself, and personally supervise the packing. If he can do this, it would probably be best for him to send them to England at once, simply wind-dried as they are, packed in a bin carefully soldered down. Even in this, however, there is some risk, as the skins must pass through the summer's heat of the plains of India. (489)

However, what is most remarkable about Darrah's *Sport in the Highlands of Kashmir* is the presence of endpapers printed to replicate the skin of a wild beast. Inside the front and back covers, a textured image of a zebra's hide replaces the traditional marbled paper. Reasserting itself as a cartographic tool, these skins (these *trompe l'œil* sheets) are shaped like envelopes or pockets into which the maps showing "The Territories of the Maharaja of Jumoo and Kashmir" and "The Great Bend of the Indus" are folded and inserted. Just as the skin of a body reveals that body's contours, so does the hide of these containers delineate the outline of the hunter's narrative.

---

accounts of the times his own skin is wounded by rhinoceros, elephants, and lions. These life-threatening encounters help map the landscape and his journey.

Figure 5.2 End paper from *Sport in the Highlands of Kashmir*
Source: Henry Zouch Darrah 1898.

The use of skin to chart foreign territories complements the practice of arranging specimens sent home to museums and zoos. Once shipped to the imperial center and displayed within either public or private places, wild skins were arranged so as to offer an imaginative spatial geography of empire. They filled in the blanks and invented rather than represented a culture.[16]

With this agendum in mind, museum curators laid out their exhibits so as to provide visitors with a tour of empire and its natural history.[17] Similarly, nineteenth-century zoos arranged displays of exotic animals and their stuffed skins to map the world. In this respect, zoo maps were essential, for how else were visitors meaningfully and authoritatively to find their way through the zoo's unfamiliar and bewildering territory? A guide, charting this alien landscape, was necessary to steer and narrate the public's course. The 1841 *Plan of the Garden of the Zoological Society within the Area of the Regent's Park* and the 1854 map of the institution published by H.C. Clarke and Co. give both a flat and an angled bird's-eye view (a privileged perspective). With fixed names and places, these maps trace the paths, label the main animal exhibitions, and offer three-dimensional drawings, featuring out-of-proportion images of visitors and caged animals.[18] Like the adventurers, explorers, and hunters traveling in uncharted lands, visitors would have also been guided (as I noted in the first chapter) by the distinctive markings of the creatures' skins as well as current classification schemes, which would function as the maps' landmarks and delineate a region's boundaries.

Zoogeography and Thematic Mapping

Animals and birds, of course, are no strangers to the cartographic idiom; they have always functioned to help orient a person's way in the world. If one glances at ancient maps, it is not unusual to find images, many implausible, of animals.[19] Creatures of the ocean and the earth have often embellished a mapmaker's depiction of a land or a sea, and have decoratively recorded the reported sightings of early explorers who sailed into uncharted regions. The inclusion of these creatures was a sign of difference. They registered the fact that the map was of an alien place, a

16 Annie Coombe is sensitive to this phenomenon. Her study *Reinventing Africa* reflects on the fact that popular ideas of Africa, for instance, were constructed in Victorian Britain through exhibitions of material and living artifacts.

17 For some thoughts on this matter, see James R. Ryan, "Hunting with the Camera: Photography, Wildlife and Colonialism in Africa" in *Animal Spaces, Beastly Places: New Geographies of Human-Animal Relations* (Philo and Wilbert 203–21).

18 These zoos and their maps are spaces placed within the firm, secure borders in cities adjacent to centers of population. Because they are set aside from the jungles, forests, and deserts, they offer merely an imitation of the colonial world. For an interesting article on zoo maps, see S. Mary, P. Benbow, and Bonnie C. Hallman's "Reading the Zoo Map: Cultural Heritage Insights from Popular Cartography."

19 For a history of animals on maps, see Wilma George's *Animals and Maps*. Also consult Arthur H. Robinson's *Early Thematic Mapping in the History of Cartography*.

region that vividly contrasted with the surroundings of home.[20] Llamas, opossum, peccaries, alligators, snakes, monkeys, dragon, large rodents, armadillos, rheas, and turtles, for instance, fill the interior of South America in Desceliers's 1546 map of the world (to be found in the John Rylands Library, Manchester) and register the foreignness of that land.

As time passed, animals as decoration or as markers of difference gave way to a more systematic depiction and study of their distribution. Zoogeographic, thematic, and distribution maps replaced the earlier ones that had sometimes marginalized these images and featured exotic oversized mammals, fish, birds, and monsters as signs of otherness in order to represent a territory outside the borders of the familiar world.

In the Victorian period, with the broadening interest in studying and organizing natural history specimens as well as with the increasing opportunities and means of travel, animals and birds were no longer either simply ornaments or symbols of difference but often became the cartographer's sole focus or subject; they were the *raison d'être* of cartography.[21] Prompted by such significant examples as Alexander von Humboldt's studies of geographical and biological distributions, thematic cartography became increasingly prominent, and by mid-century thematic zoological maps were as visible as their botanical equivalents, which had preceded them (see Robinson 60–62). As cultural products, these figuratively encoded the nineteenth-century preoccupation with ordering, arranging, controlling, and displaying the natural world—of putting everything in its place. They each dealt with a single category of information (for instance, birds and reptiles) and focused on variations from place to place rather than on individual locations. This orientation reflected Alfred Wallace's field observations as well as Charles Darwin's interest, during the voyage of the *Beagle*, in noting the distribution of the wildlife he saw. As Philip J. Darlington, Jr. observes in *Zoogeography: The Geographical Distribution of Animals*, determining the distribution of animals in relatively undisturbed parts of the world was "one of the proofs of evolution" (14).

One of the most vivid nineteenth-century maps which depicts the earth's skin exclusively through the presence of animals is the *Zoological Map of the World, Shewing the Geographical Distribution of Animals*, published in London by

---

20 Wilma George notes: "For the most part it was the large animals … that came to the notice of the explorers and were noted in the journals. It was these animals that found their way on to the maps. The tendency was mainly to stress the differences between the various parts of the world rather than similarities and the abundant large animals provided a more striking contrast with home" (25).

21 Animals were not the first subjects of thematic maps. Zoogeography followed the numerous attempts to replicate the pattern of dynamic geographical phenomena, such as the circulation of oceans and the direction of winds, as well as the distribution of plants or vegetation.

William Spencer on November 5, 1845.[22] Lively colored images of beasts, birds, and fish take every available space so as to capture the complexion and character of each continent, island, and ocean. The living creatures, roaming the earth and inhabiting the oceans, fashion the world's exterior and form its envelope so that the world dons the hides of beasts. (It is interesting that many of the scenes depict these animals being hunted for their skins.) Creatures, which fly through the sky, race across central plains, swing from trees, run in packs, and rest on rocks, plot, mark, and cover the earth's surface. At times they seem to rule the world: the whale tips the whalers' boats; the bear attacks a lone man; a crocodile pursues a native; and a boa strangles an unsuspecting boatman. At other moments, as if representing published travel narratives, hunters with spears and guns shoot and wrestle with these wild beings. What is essential is that in this map these creatures are the defining and active presence of the world.

Due to the increasing attention to zoogeography, by the mid-nineteenth century mammals, reptiles, birds, and fish were functioning more actively than ever to represent the earth's surface, the skin of the world. (Of course, there were also other categories of maps.) Coordinating collected data concerning these creatures with newly devised cartographic symbols as well as taking advantage of new printing methods (such as lithography), maps were dedicated exclusively to charting the territory of a certain creature's existence and showing the range of single species of a genus. W. and A.K. Johnston's 1848 *The Physical Atlas*, for instance, includes a world map showing the distribution of snakes, as well as the distribution of monkeys and lemurs.[23] The limits of these animals' territories were of more concern or interest than the already measured coastline or surveyed border of a country or continent.

To make a distribution map, it was assumed that the shape of the world's surface and land masses were already precisely determined and secure; what needed to be done was to superimpose the animal territories upon that structure in order to divide the globe into zoogeographical regions. The belief was that "every zone, every climate, nearly every country, has a variety of plants and animals more or less peculiar to itself" (*Johnston's Geographical HandBooks* 18–19). Two examples are the *Zoological Map Showing the Distribution of Animals over the*

---

22 A list of the animals exhibited on *Zoological Map of the World, Shewing the Geographical Distribution of Animals* follows (the spelling of these species is that found on the map): Australia: Black Swans, Parrots & Cockatoos, Kangaroos, duck Bill, Lyre Birds; Vandiemens Land: Kangaroo Rat; New Zealand: Amterix; Africa: Ant Eater, Ostrich, Lion, Water Buffalo, Elephants, Hippopotamus, Crocodile, Rhinoceros, Antelope, Pelican, Jackal; India: Elephant, Wild Boars, Tiger, Boa Constrictor; Borneo: Orang Outang; Arabia: Arabian war horses, Dromedaries, Apes; Asia: White Bears, Brown Bears; South America: Wild Cattle, Armadillos, Boa Constrictor, Chinchilla, Ficogna, Rheas, Jaguar, Monkeys, Parrots, Macaws, Toucans, Llamas, Ring Vultures, Condor, Harpy Eagle, Ant Eaters, Sloth, Puma; Ocean: Fish, Auks, and Flying Swans (Maps 950 [7]).

23 For reproductions of these maps, see Arthur H. Robinson's *Early Thematic Mapping in the History of Cartography*, 106–7.

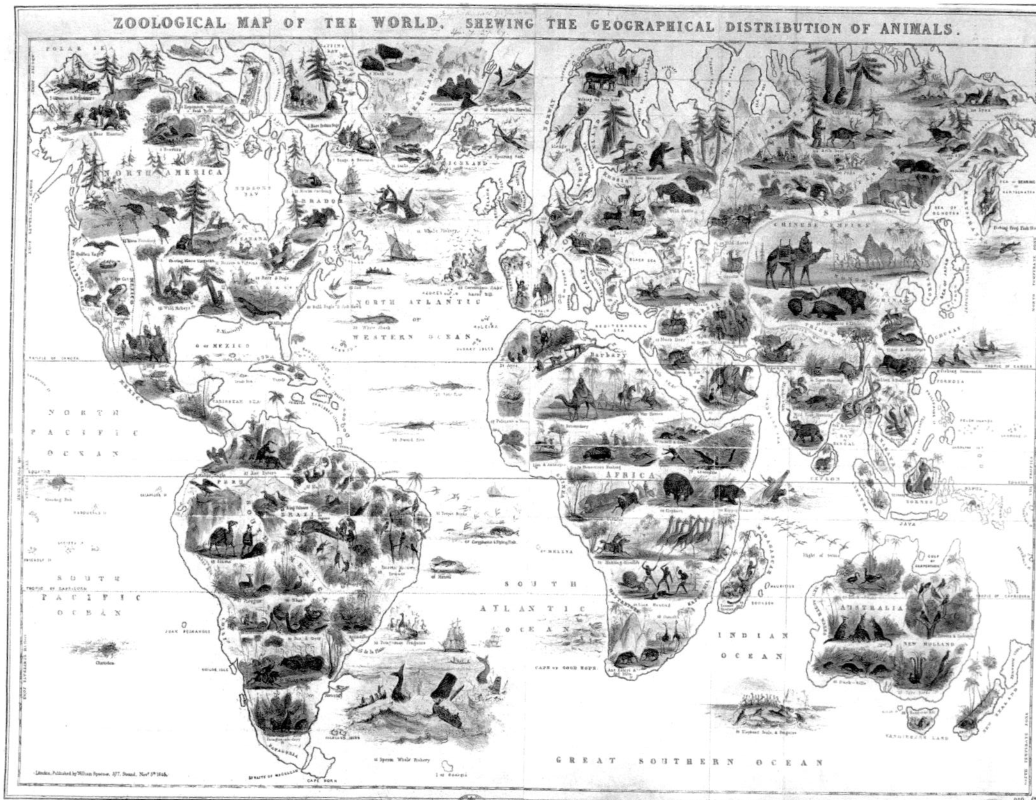

Figure 5.3 Zoological Map of the World, Shewing the Geographical Distribution of Animals
Source: British Library Board.

World and the *Zoological Map Showing the Distribution of Birds & Reptiles over the World* printed in *Physical Map of the World*, published by James Reynolds in 1849.[24] Note the lines that cut across and circle portions of the global image so as to indicate the various districts in which certain animals prevail. Delimiting the boundaries of a creature's habitation at a glance was an important part of this kind of map. *The Zoological Map Showing the Distribution of Birds & Reptiles over the World*, for instance, attaches the following explanation:

> Birds occupy the whole globe. Voyagers in high latitudes have never gone beyond their horizontal range, while the habitat of the Condor extends perpendicularly from the plains to the loftiest crest of the Andes. Reptiles occupy the lowest rank among terrestrial vertebrae, they diminish in number, magnitude, and noxiousness from the equator to the poles. (British Library Maps 950 [33])

Perhaps the most famous of these maps depicting the geographical distribution of animals was that designed by Wallace, who, as we have seen in the second chapter, was one of the most prolific collectors of skins and natural history specimens. Rather than attempt to compose a separate "coloured or shaded map for each family" (the number of maps would have been unwieldy and costly),[25] Wallace favored the development of more general views and divided the world into six regions: Palearctic, Nearctic, Neotropical, Ethiopian, Oriental, and Australian (*The Geographical Distribution of Animals* ix). Although he recognized conceptual weaknesses in the structure, he generally subscribed to the belief that each region contained distinctive animal life and was separated from other areas by obvious geographical features, such as an ocean or a mountain range.[26] These maps were yet one more attempt to schematize, and thereby to manage, an ever-increasing knowledge of the world's flora and fauna. In a sense, they too belonged to the colonial culture's impulse not only to learn more about areas distant from the metropole but also to gain some sort of mastery over those regions.

---

24 It is somewhat of a privilege to have seen these maps, for, as Arthur H. Robinson points out, many of the thematic maps made after the beginning of the nineteenth century, "especially those in journals and inexpensive atlases," are on paper that is rapidly deteriorating (xi).

25 Alfred Wallace, though, did occasionally make separate maps. In 1860, for instance, he published an essay about the "Zoological Geography of the Malay Archipelago" in the *Journal of Proceedings of the Linnaean Society*, and in 1863 he read a paper on the same subject before the Royal Geographical Society.

26 The Neotropical region, for instance, covered the whole of South America, most of Mexico, and the West Indies. It was joined to the Nearctic by the Central American isthmus and separated from all other regions by the sea. It was assumed that no animal in that area would cross such a boundary (think of Darwin's finches on the various islands of the Galapagos).

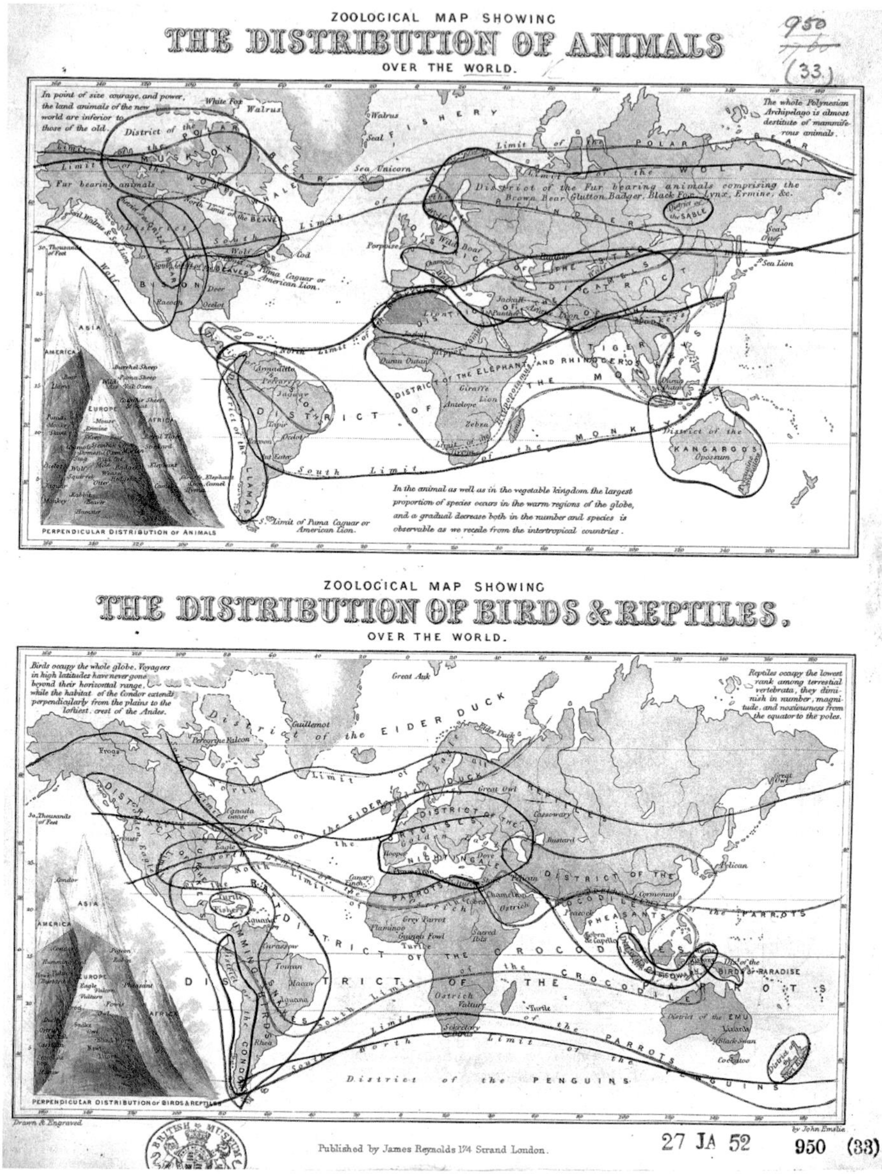

Figure 5.4 Zoological Map Showing the Distribution of Animals over the World, and Zoological Map Showing the Distribution of Birds & Reptiles over the World

Source: British Library Board 1849.

Part Three: Gerard Manley Hopkins: Animals in the Landscape

Perhaps surprisingly, one writer whose cartographic imagination was significantly affected by zoogeography and the distribution maps discussed in the preceding section, is Gerard Manley Hopkins. This third part of the discussion concerning animals, skins, and the mapping of the Victorian landscape not only demonstrates Hopkins's awareness of the orientation of these maps but also illustrates the ways in which he converted their static and fixed perspective so as to proffer a more animistic view of nature and honor his engagement with a sensuous geography. To put it plainly, Hopkins used this form of cartography as a basis with which to create a different kind of map.

Hopkins and the Zoogeographic Tradition

Well-established and popular, mid-nineteenth-century thematic maps displaying the distribution of animals were, as I have implied, part of the public's consciousness and therefore contributed to the way many people conceived of the world outside their immediate boundaries.[27] Hopkins's sensitivity to this cartographic form is understandable, for both he and his immediate family were intensely attracted to a nineteenth-century culture preoccupied with, if not devoted to, the study of zoology, botany, physics, chemistry, and geology.[28] Indeed, Hopkins's father (Manley Hopkins) went so far as to publish an essay, "Essex Elephants," in an 1860 issue of *Once A Week: An Illustrated Miscellany of Literature, Art, Science & Popular Information*. (The magazine's subtitle emphasizes the appeal of its subjects to the general public.)[29] Appropriately, not only does this article convey the father's as well as the period's avid interest in geology, but more significantly articulates Manley's excitement when an area close to where they lived was excavated and revealed the bones of animals that had once inhabited and mapped the landscape. In the style of the mid-century thematic and distribution maps, these animal remains, such as the immense mammoth tusk 14 feet below the level of the

---

[27] Such documents were not confined to scientific institutions and publications; they also, for instance, hung in schoolrooms, offices, or businesses and were readily available in cheaply published atlases. The period's remarkable interest in natural history also encouraged this orientation.

[28] Critics such as Marie Banfield, Gillian Beer, Jason H. Lindquist, Norman Mackenzie, Julie V. Nixon, and Tom Zaniello have within the last decade recognized and, in varying degrees, explored Hopkins's dialogue with physics, chemistry, and natural history. They have firmly established the scientific context of Hopkins's work. Norman Mackenzie, for example, remarks on Hopkins's father's (Manley Hopkins's) interest in the collecting of fossils and discovery of animal bones ("Hopkins and Science" 87).

[29] For an excellent discussion of Manley Hopkins's various interests and his contribution to the development of Gerard Manley Hopkins's attention to art and painting, see the first chapter, "Early Influences," in Catherine Phillips's *Gerard Manley Hopkins and the Victorian Visual World*.

soil at an Ilford brick-field, mark and plot the landscape of an older time. Manley is thrilled not just because of this one find but also because there are similar fields throughout Britain. The excavation at Ilford is not the only place where these vestiges lie. In other areas, uncovered bones serve as landmarks of a larger map. As he explains:

> Though the excavations at Ilford have been singularly productive in the discovery of animal remains, it is not to be understood that they exist in that site only. In other parts of Essex and also in Middlesex coming within the basin of the Thames, similar bones have been brought to light. Remains of the elephant have been met with at Grays, at Harwick, at Erith, at Brentford, at Kingland, and within a few months past, at Charing Cross. At Erith the lion and hyeana, [*sic*] and at Grays the bear, add the carnivora order to the list of animals given above. (*Once A Week* 3: 55)

As if prompted by his father's keen interest in these fossilized remnants, Hopkins periodically seems to have called on zoogeography in order to map a landscape and capture the character of the earth's surface. As a result, throughout his diaries, journals, and sketchbooks, images of animals emerge as significant cartographic tools. They are the raw materials of Hopkins's vision.[30]

The most striking example of Hopkins's use of animals and their skins to map his surroundings comes from a July 15, 1868 journal entry in which Hopkins describes a glacier he admires while on an Alpine holiday with his university friend Edward Bond.[31] Wanting to capture the glacier's essential outline (and perhaps its formidable presence), Hopkins summons the timely image of a white tiger's hide. (The hide he has in mind is probably a colonial trophy which has been preserved, perhaps by Ward's firm, and shaped into a rug.) The metaphor of this wild animal's skin maps the glacier. In the following passage, Hopkins explores the similarities between the texture and the shape of the hide and the outward appearance of the

---

30 Even though it is possible to find such figures in his poetry—consider "The Windhover," for instance—in this discussion, I shall concentrate on these notebooks, for it is within their pages that Hopkins consciously and fervently labored to map his surroundings so as to catch the organization of its surfaces.

31 On July 3, 1868, Gerard Manley Hopkins, then in his middle twenties, started for Switzerland with Edward Bond, a university friend. This was to be Hopkins's last holiday before he entered Manresa House and Stonyhurst as a Jesuit novitiate. He and Bond remained in the Alps until the end of the month. (For a full description of Hopkins's itinerary in Switzerland, see Tom Zaniello's "Alpine Art and Science: Hopkins's Swiss Adventure," 3–17.) During those several weeks, the two of them, with a Baedeker in hand, followed the familiar routes and, like many other visitors, went up to such places as the Rigi and visited the glaciers at Grindelwald. (See also Colley, *Victorians in the Mountains: Sinking the Sublime*, 173–5).

glacier—correspondences that are made even more vibrant because they exist in a context of difference.[32]

> There are round one of the heights of the Jungfrau two ends or falls of a glacier. If you took the skin of a white tiger or the deep fell of some other animal and swung it tossing high in the air and then cast it out before you it would fall and so clasp and lap round anything in its way just as this glacier does and the fleece would part in the same rifts: you must suppose a lazuli under-flix to appear. The spraying out of one end I tried to catch but it would have taken hours: it was this which first made me think of a tiger-skin, and it ends in tongues and points like the tail and claws: indeed the ends of the glaciers are knotted or knuckled like talons. Above, in a plane nearly parallel to the eye, becoming thus foreshortened, it forms saddle-curves with dips and swells. (*Journals* 174)[33]

Hopkins's pencil drawing of this glacier in "Sketchbook C" (containing sketches made between August 16, 1866 and July 24, 1868) outlines the linear scheme of the glacier's surface and simultaneously reveals the pattern and shape of the tiger's splayed hide. The metaphor's tenor and vehicle reside in the same frame; in the illustration they combine to form a two-dimensional map.

In other parts of his journals, Hopkins periodically returns to comparable metaphors. For instance, on March 14, 1871 Hopkins gazes at the Pendle hills covered in snow and spontaneously notices how the ruts in that land, "plated by snow," resemble the patterned markings of a zebra's hide. Once more a wild skin inhabits and maps the earth's surface:

> Pendle was beautiful: the face of snow on it and the tracks or gullies which streaked and parted this well shaped out its roundness and boss and marked the slow tune of its long shoulder. One time it lay above a near hill of green field which, with the lands in it lined and plated by snow, was striped like a zebra: this Pendle repeated finer and dimmer. (*Journals* 205)

As if recalling the zoogeographic idiom, throughout his journals Hopkins continues to summon these wild skins in order to plot the defining lines of the land, the sea, and the sky. The image of "many silvery worms" moving "to and from one another" helps Hopkins describe the wimpling patterns of the blowing

---

32 In a notebook, dated May 12, 1865, Hopkins responds to a discussion on Plato's concept of the origin of beauty. In the discussion, he identifies the irregularities of nature as a source of beauty and speaks of the importance of "likeness and difference or agreement and disagreement or consistency and variety or symmetry and change" (*Journals* 90).

33 All references to Hopkins's notebooks, diaries, and journals come from Humphrey House's edition, *The Journals and Papers of Gerard Manley Hopkins*. Subsequent quotations from this text are marked as *Journals*. Important work on these materials has been done by Norman Mackenzie and is being continued by Lesley Higgins of York University.

Figure 5.5 "This is a glacier, though you wd. not think it. July 13, '68"
Source: Balliol College, Oxford.

fine snow, gliding over the ground (196); the form and color of a snake's shed skin convey the "inscape" of the bluebells standing in a "little wood" (208), and "tufts of ermine" structure the clouds (243). Moreover, on February 24, 1873, "a pale goldfish skin without a body" (236) profiles the prospect of a ridge illuminated by a bright sunset. For Hopkins, all these creatures (e.g. the tiger, the zebra, the worms, and the snake) are part and parcel of the world; they are embedded in its surface.

Although Hopkins never explicitly acknowledged that he was beholden to such publications as the 1846 *Zoological Map of the World, Shewing the Geographical Distribution of Animals*, the above instances from his diaries and journals suggest that he evoked, figuratively, and occasionally literally, the genre's example. For instance, in his early diaries (1862–1866), Hopkins mentions that when walking with his friend Addis in the fields near Oxford, he saw "a snake glide through a hedge" (*Journals* 22). Attentive to the twigs bent by the snake's curving passage, Hopkins, as cartographer, records the scene's defining lines. In an accompanying sketch, he follows the snake's twisting motion and fixes the contours of that patch of earth. The snake has literally mapped this landscape.

Figure 5.6 Sketch accompanying Sunday, April 10 (1864) entry in Hopkins's *Journals*
Source: Journals 1864.

Not surprisingly, however, unlike the distribution maps, animals and their skins, as in the example of the white tiger's and the zebra's skins, generally play a metaphoric, rather than a literal role. These figures help Hopkins chart the textures and shapes of the landscape. As a result, dispersed throughout Hopkins's early diary and journals are passages about the "Moonlight hanging or dropping on treetops like blue cobweb" (23); "thick-fleeced bushes" which resemble a "heifer's ear" (38); "Feathery rows of young corn" (57); clouds which resemble "the ringlets of a ram's fleece blowing away," and a sky shaded into "tortoise-shell squares" (212).

As if also recalling the phenomenon of the excavated animal remains under Ilford, Erith, and Charing Cross, Hopkins periodically focuses on these scattered pieces of creatures as they seem to map the texture of a landscape. Like the yellow feathers from a "goldencrested wren," which once got into Hopkins's room at night, these fragments capture the texture and contours of a moment (*Journals* 227).[34] Elsewhere in his journals, skins or bits and pieces of animals are dispersed

[34] The passage reads: "A goldencrested wren had got into my room at night and circled round dazzled by the gaslight on the white ceiling; when caught even and put out it would come in again. Ruffling the crest which is mounted over the crown and eyes like bettlebrows, I smoothed and fingered the little orange and yellow feathers which are hidden

via metaphors throughout the landscape. Trees have a rough "furry touch," (77), and beech leaves are covered with a "silver fur" (135). Clouds are compared to flocks of sheep, to cobwebs, to feathers, to "silver bright" fish scales (147), to schools of mackerel, and to a mare's tail (138, 150, 161). A tree silhouetted against the light displays "antler-like sprays" (149); the leaves of an ash tree by the side of a gushing river in the Alps divide like a "bat's wing" (177); a waterfall gives off showers of drops which are "strung together into little quills" (178); long icicles bend over and curl "like so many tusks and horns" (201), and bluebells "look like the heads of snakes" (209). A limb of an Alpine glacier is shaped "like the fan-fin of a dolphin" (178); the feeder glacier from Monte Rosa resembles "a turbot's tail" (181), and the waves of the sea rise "like fishes' backs" (184).

Hopkins and the Idea of Mapping

That Hopkins would have an inclination to map the landscape at all is not only due to the zoogeographical tradition and the general cultural interest in arranging and ordering the natural world, but also indebted to his own cartographic impulse. To those familiar with Hopkins's work, this inclination might be disorienting, for critics more frequently discuss the considerable effect of the fine arts on his descriptions of landscape. For instance, Catherine Phillips, in her excellent study of the influence of the Victorian visual world on Hopkins's writings and drawings, thoroughly discusses not only Hopkins's keen interest in and knowledge of painting but also the pleasure he took, in his own work, especially when considering shades of coloring as well as a scene's perspective.[35] Although I concur with this point of view, I suggest that another aspect of that "visual world"—the map—also played a significant role in his understanding of his surroundings.

Before continuing with a discussion of Hopkins's distinctive use of animals in the mapping of the landscape, I think it helpful to digress (but only seemingly) from the immediate concern and consider why, in the first place, Hopkins was so drawn to the image of a map. The digression is lengthy, but necessary if one is more fully to understand why Hopkins periodically elicited images of animals and their skins when forming his maps.

As commentators recognize, Hopkins diligently, if not compulsively, searched for what he labels the "inscape," the "scaping," or the "law" of each scene. He was intent on actualizing his conviction that "all the world is full of inscape and [that] chance left free to act falls into an order as well as purpose" (*Journals* 230).[36] Because of this quest to isolate essential patterns in things, the literal and

---

in it. Next morning I found many of these about the room and enclosed them in a letter to Cyril on his wedding day" (*Journals* 227).

35 For detailed discussions of the importance of the fine arts for Hopkins, see Phillips's *Gerard Manley Hopkins and the Victorian Visual World*.

36 The animal metaphors enumerated above are integral to this impulse to map the inscape of a thing. His memory of the shape of a bat's wing or the pattern of a zebra's skin

emblematic figure of a map, I suggest, was particularly meaningful, for it offered Hopkins a viable structure through which to register this "scaping," and also proffered him a means of inscribing and fixing the uniqueness of a tree, a river, a waterfall, a mountain, or a leaf.[37] His cartographic imagination depended upon this pattern making. Moreover, this orientation contributed to a form of stability, for within the map's frame, Hopkins understood that once discovered, neither motion nor decay could disfigure the "law." For instance, when he watched the wind toss a group of chestnut trees and cause them to plunge and cross one another "with a deeper and deeper stoup," he knew that their inscape (their defining pattern) would not be lost in the disorder of that motion (*Journals* 199). And when old skin grew slack and decayed ("the pining of the skin"), he understood that the "inscape" lingered.[38] Even in a wilting flower, its beautiful inscape remained (211).

Pages throughout Hopkins's sketchbooks and diaries reveal that the mapping eye has slipped in and filtered his topographical vision so that he might catch the "law" of his surroundings. The desire to map what is before him is a driving force.[39] Because of this orientation, Hopkins does not fill in a scene's particulars so that each part blends through the passage of its shadow into its neighboring segment. Rather, in the manner of a cartographer and with the assistance of metaphor, he isolates the few defining or leading lines both verbally and visually, which locate the subject and steady the scene's undulating forms.[40] Passages from his journal capture many

---

helped him capture the essence or "scaping" of what his eyes encountered.

At the October 2013 North American Victorian Studies Association Conference, I listened to Simon Reader's insightful paper on Hopkins's impulse for pattern making. The paper was entitled "'Among the Loose Papers': Gerard Manley Hopkins's Idle Documents." Reader, from the University of Toronto, is working on Hopkins's diaries and notebooks.

37 Among Hopkins's few remaining early papers, for instance, are a number of maps he either traced or copied as a young boy. Childhood maps of Eastern Europe, two ancient Italian provinces, Campania and Lucania, and of Burma, Laos, and Anan are scattered among his papers in the Bodleian Library, Oxford. Remarkably, these survived when so little else did. Later, when he was teaching classics, Hopkins not only studied ancient maps but also, as was the style in teaching classics, used them to illustrate his lessons.

38 "It is not that inscape does not govern the behaviour of things in slack and decay as one can see even on the pining of the skin in the old" (*Journals* 211).

39 As several critics have pointed out, Hopkins not only used the concept of inscape when writing in his journals but also when structuring his poetry. In a letter to Robert Bridges, Hopkins once declared "as air, melody, is what strikes me most of all in music and design in painting, so design, pattern or what I am in the habit of calling 'inscape' is what I above all aim at in poetry" (February 15, 1879, l. 66). See Phillips (35) and Colley, "Gerard Manley Hopkins and the Idea of Mapping" (523–8).

40 Hopkins does not submit to the examples of John Ruskin's struggles to analyze each section, twist, hollow, and detail of the perspective before him. Nor does he feel compelled to account for every moment of the eye's work. Instead he prefers to pull away from the infinite regress of Ruskin's minutiae and chart the lines which delineate the edges and internal boundaries of the scene's form. Although Hopkins admitted that he sketched in a "Ruskinese" style he was not intent on rendering polished drawings. Compare, for instance,

such moments. As if mapping the sky, he searches for its essential lines: “The map of the sky was a rhomboid of grey round-moulded cloud in one great cloth stretching over the sky with one part resting somewhere on the skyline in the S. or S.W.” (*Journals* 138–9). A comparable perspective is at work when he records seeing delicate clouds forming a pattern of angled pieces (168), or when he catches the “spraying” of the thick fir and beech woods on a mountain side (171). As if holding a compass, Hopkins maps each scene’s intersecting paths. Similarly, he charts the surface of water by isolating the lines which define or wrinkle the skin of its face. And by noting the clefts that run “deep through the ice markedly in planes” (202), he diagrams an icy pond. He also traces the cracks on a frozen tadpole basin. At times Hopkins even consciously recalls the discriminating structure of relief maps. For example, in February 1873, he explicitly compares the prospect of snow-covered hillocks to the projected image of a relief map: “In the snow first-topped hillocks and shoulder outlined with wavy edges, ridge below ridge, very like the grain of wood in line and in projection like relief maps” (230).

Figure 5.7 Sketch accompanying July 2 (1868) entry in Hopkins’s *Journals*
Source: Journals 1868.

Figure 5.8 Sketch accompanying July 10 (1868) entry in Hopkins’s *Journals*
Source: Journals 1868.

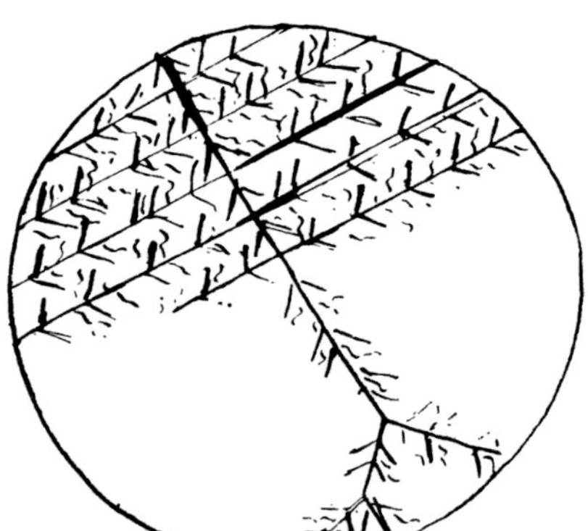

Figure 5.9 “Ice on my tadpole basin formed as below”
Source: Journals (March 28) 1868.

Hopkins’s description of the ice and its effect upon the soil in his garden with Ruskin’s. See *Journals* 201 and the opening of the “Burma Artifex” chapter in Ruskin’s *Deucalion.*

Beholden to conventional cartography, Hopkins, at times, appears (as did the windhover) to hover above the scene so as better to grasp its outline.[41] (In a sense, he sometimes proffers a bird's-eye view.) In July 1866, for instance, he maps the landscape beyond the town of Midhurst: "Just beyond the town … runs the canal water looking like a river and on the steeper-rising further side the park trees make a towering and noble wall which runs along to the left and turns and embays a quarter of a mile away, the whole having the blocky short cresting which freely grown park trees show" (*Journals* 145). Subscribing to the cartographer's imagination, Hopkins also fixes a prospect by isolating and identifying its landmarks. When his eye sweeps over the view from Cumnor Hill (133–4), for instance, Hopkins reverts to conventional markers, such as a church, a road, and a bridge, and intermingles these with less-traditional points of orientation, such as the sky, trees, hedges, and birds. These signposts hold the landscape still and outline the scene's "law."[42]

Like a cartographer Hopkins is also conscious of internal boundaries—the borders—which separate one entity from another. He heeds and presses toward the edges of all he sees. This consciousness of boundary is apparent from the beginning of his journals. For instance, on May 10, 1866 he wrote about the straight line that separates one part of the sky from another. On the same day he recorded that he had seen children "with white rods beating bounds of St. Michael's parish" (136)—their defining movements recall the ritual of the women's walking club in *Tess of the D'Urbervilles*. Hopkins's tendency to notice edges persists throughout his journals. He is aware that the hill is "crowned" with beautiful trees (140), that the common is "edged" with a frieze of trees (154), that the oaks are "freshly edged" (157), and that the grain's "edges" are "sharpened" (168). A drawing in

---

[41] Hopkins's impulse to map his surroundings once even extended to an evening at the theater, in which he mapped the "visible law" of the crowd attending the production (see *Journals* 139).

[42] From Cumnor Hill saw St. Philip's and the other spires through blue haze rising pale in a pink light. On further side of the Witney road hills, just fleeced with grain or other green growth, by their dips and waves foreshortened here and there and so differenced in brightness and opacity the green on them, with delicate effect. On the left, brow of the near hill glistening with very bright newly turned sods and a scarf of vivid green slanting away beyond the skyline, against which the clouds shewed the slightest tinge of rose or purple. Copses in grey-red or grey-yellow—the tinges forerunning the opening of full leaf. Meadows skirting Seven-bridge road voluptuous green. Some oaks are out in small leaf. Ashes not out, only tufted with their fringy blooms. Hedges springing richly. Elms in small leaf, with more or less opacity. White poplars most beautiful in small grey crisp spray-like leaf. Cowslips capriciously colouring meadows in creamy drifts. Bluebells, purple orchis. Over the green water of the river passing the slums of the Town and under its bridges swallows shooting, blue and purple above and shewing the amber-tinged breasts reflected in the water, their flight unsteady with wagging wings and leaning first to one side then the other. (*Journals* 134–5)

"Sketchbook D" showing the outline of a tree captures this perspective.[43] That these boundaries, lines, and edges are important to Hopkins is later evident in his poems, such as "Pied Beauty" which celebrates the landscape that is "plotted and pieced" (l. 5), or "God's Grandeur," which mourns the loss of these borders and laments that "all is seared with trade; bleared, smeared with toil" (l. 6).

Figure 5.10 "Sept 4, '68" from Sketchbook D
Source: Balliol College, Oxford.

A Different Sort of Map

Although Hopkins was often drawn to the conventions of traditional cartography discussed above, he was not exclusively beholden to their requirements. Various journal entries and accompanying sketches that appear to replicate the flattened miniaturized perspective of the two-dimensional map, as well as those map-like figures which trace an area's defining lines (see, for example, Figure 5.10), do not, however, fully represent his cartographic imagination. Hopkins also embraced a different sort of map—one which admitted motion and invited the body to haunt

43 For a detailed discussion of Hopkins and the idea of mapping in his poetry, see Colley, *Victorians in the Mountains: Sinking the Sublime.*

and touch the contours of the earth's surface. Evoking and transforming the ancient perspective of the corporeal earth, Hopkins, throughout his notebooks and journals, constructs a sensuous geography in which the moving body mediates the mapping. As we shall see once the discussion returns to its focus to the chapter's primary subject, Hopkins's use of animals and wild skins in the mapping of the landscape was fundamental to this understanding. First, though, it is helpful to examine this different kind of map.

One characteristic is that the landscape is seen from the viewpoint of one moving through it.[44] Hopkins's cartographic imagination refuses to exclusively acknowledge the impersonal, abstracted, or timeless eye of the static zoogeographic map; instead, the eye is also allowed to chart the movement and fluctuations of the things it passes by. In the following passage, for example, Hopkins maps the land by tracking his progress as he finds his way along a lane in view of the Wye hills. The perspective belongs to the "I" who is walking through this landscape, from landmark to landmark. Hopkins, consequently, notes the "contradictory supple curvings in the boughs" and recognizes that the sheep are, just for a moment, grazing by "the rise of the lane." They will move on. Unlike the eternally stationary and enclosed creatures depicted on the 1845 *Zoological Map of the World* (see Figure 5.3), these sheep belong to the rhythm and progressive activity of Hopkins's shifting eye, which looks first to one side "and then" to another. The map acknowledges that they, like the observer, are subjects of time.

> The afternoon way we much enjoyed, in especial we [Hopkins and Addis] turned down a grass lane to reach the river at the ferry. It was steep down at first and I remember blue sprays of wych-elm or hazel against the sunlight green further on. Then the fields rose high on each side, one crowned with beautiful trees (there was particularly an ash with you could not tell how many contradictory supple curvings in the boughs), and then orchards, of which this country is full; on the other, with a narrow plot of orchard in which sheep grazed between the rise and the lane, was Goderich castle of red sandstone on the height. (*Journals* 140)

In many other entries, Hopkins's frequent use of the participle helps keep the map in motion so that one is aware of a vibrant, shimmering landscape rather than a bounded area frozen in place: the moonlight is "hanging and dropping on treetops" (23); drops of rain are "hanging on rails" (72); and clouds are "growing in beauty" (138). This different kind of map represents particulars in a state of becoming and receding. However, this continuous motion does not, in Hopkins's mind, obliterate his sense of the landscape's inscape, for, as I have already explained, once he has discovered the "law" of the landscape, he can safely step into the shifting motions of his surroundings as well as into the spaces in between. He is free to

---

44 Phillips is sensitive to Hopkins's incorporation of movement in his understanding of inscape. She remarks: "he defined the inscape of the windhover from its characteristic movement as much as from its physical appearance" (53).

engage a different kind of map in which he can still honor the surface's inscape but also recognize its fluctuations. He can place his body into the flux of a landscape without fear of being lost.[45]

What makes Hopkins's mapping of his surroundings, either through his metaphors or his sketches, so engaging is the fact that, contrary to the sterile, abstracted outlines of conventional maps, they recognize the body's sensuous encounters within a landscape—a perspective that he shares with Hardy, though to a more intense and complicated degree. To put it more plainly, in Hopkins's cartographic imagination, the body inhabits the landscape. Indeed, for him the world itself is a body (whether that be a human or a vibrant, living creature). As Phillips observes, Hopkins often "harnessed" animal/human images of limbs and shoulders in his descriptions (2). From Hopkins's perspective, the landscape has fingers, knuckles, earlobes, eyes, eyelids, nails, hairs, limbs, ribs, tongues, vertebrae, heads, hands, foreheads, shoulders, veins, toes, and lips. Drops of rain are like fingers; trees have lobes; clouds have spines; the sky displays a "burnished" forehead (199); and the bluebells "rise to the end like a tongue" (209). Moreover, "the sea foots or toes the shore and inlet" (221), and "heads of hair" tuft the grass (228).[46]

Because of this sensitivity to bodily presence throughout his prose, Hopkins envelopes the landscape in skin and speaks of a sky that "blushes" (*Journals* 154). He attends to the earth's bones, "sleeved in flesh" (72)—a metaphor which faintly resembles passages describing the landscape in *Tess of the D'Urbervilles*, but more directly recalls one of Mary Wollstonecraft's published letters in which she compares the rude Norwegian sea to "the bones of the world waiting to be clothed" (28). And he remarks, for instance, on the oak trees, which spread their "solid and muscular" roots underneath the earth's skin. Occasionally, his eye focuses on such images as sheets and canvases, which stretch "over the sky" to form a kind of skin

---

45 In a recent talk at The Victorian Tactile Imagination Conference held at Birkbeck College, July 19–20, 2013, William A. Cohen spoke of the tactile mode in Thomas Hardy's *Woodlanders*, which disperses agency among human and nonhuman elements in this novel's landscape. See also Cohen's discussion of Thomas Hardy's *The Return of the Native* in *Embodied: Victorian Literature and the Senses,* 86–107.

46 One entry (January 23, 1866) from his early diaries shows Hopkins, almost too diligently, expressing his sense of the body's presence:

> Drops of rain hanging on rails etc seen with only the lower rim lighted like nails (of fingers). Screws of brooks and twines. Soft chalky look with more shadow middles of the globes of cloud on a night with a moon faint or concealed. Mealy clouds with a not brilliant moon. Blunt buds of the ash. Pencil buds of the beech. Lobes of the trees. Cups of the eyes. Gathering back the lightly hinged eyelids.
>
> Bows of the eyelids. Pencil of eyelashes. Juices of the eyeball. Eyelids like leaves, petals, caps, tufted hats, handkerchiefs, sleeves, gloves. Also of the bones sleeved in flesh. Juices of the sunrise. Joins and veins of the same. Vermillion look of the hand held against a candle with the darker parts as the middles of the fingers and especially the knuckles covered with ash. (*Journals* 72)

on which to map a prospect (*Journals* 138). Sometimes the ground streaked with snow is "sheeted" like a taut skin (228).

Skin not only maps the landscape's complexion or exposes its muscular undertones but also reveals the scars of the earth. As if paralleling Hardy's sensitivity to Tess's wounded flesh and her injured landscape, Hopkins heeds the "deep flesh cuts" in the surface of a glacier (at Grindelwald) "where one sees the blood [of the earth's body] flush and welling up" (175). A day or so later, when glancing at the clouds, he notices their "shapeless skins of fretted make" (179).

Believing that the body and particularly the skin "gives depth and projection to the landscape" (*Journals* 168), Hopkins replaces the cartographer's disinterested and distant eye with a sensuous body that brushes up against and moves through its surroundings.[47] The senses reach out like the limbs of the Renaissance man splayed across the breadth of the cosmos. As a result, in his journals Hopkins maps what he sees from the prospect of someone who captures the scaping of a rock pool by "plunging" his feet in it, by noticing the oil left on his skin, and feeling the "texture" of the pool's edges (224). Like Tess, Hopkins is actually part of the landscape's flesh. He too wanders within and is touched by its skin's envelope. He "tumbles over the moraine of a glacier" (181), feels the smoothness of the oak tree's roots, and senses the thickness of the furze on Hay Tor (155). His is a haptic landscape. In other journal passages, the experience of touch is significantly integral to Hopkins's ability to grasp the pattern of what he sees. Indeed, on one afternoon in late August 1867, Hopkins literally puts his hand (the primary organ of touch) up to the sky and spreads his fingers so that their flesh, as well as their figurative reach, might catch the defining light and map the clouds' colors. Though the hand is not literally touching the sky, it extends itself through the idea of touch to become the medium through which Hopkins organizes the spatial territory and identifies the hue of the sky above him:

> Fair; in the afternoon fine; the clouds had a good deal of crisping and mottling.—A round by Plumley.—Stands of ash in a copse: they consisted of two or three rods most gracefully leaved, for each wing or comb finally curled inwards, that is upwards.—Putting my hand up against the sky whilst we lay on the grass I saw more richness and beauty in the blue than I had known of before, not brilliance but glow and colour. It was not transparent and sapphire-like but turquoise-like, swarming and blushing round the edge of the hand and in the pieces clipped in by the fingers, the flesh being sometimes sunlit, sometimes glassy with reflected light, sometimes lightly shadowed in that violet one makes with cobalt and Indian red. (154)

---

47 Paul Rodaway in *Sensuous Geographies: Body, Sense and Place* speaks of "haptic geographies": "Touch geographies are the sensuous geographies arising out of the tactile receptivity of the body, specifically the skin, and are closely linked to the ability of the body to move through the environment and pick up and manipulate objects" (41).

Animals, Inscape, and Sensuous Geography

With a better sense of Hopkins's attraction to the cartographic idiom as well as an understanding of his appreciation for a sensuous geography, it is possible to comprehend more fully why Hopkins was drawn to the images of animals and skins in the mapping of the landscape. These images (e.g., the flung tiger skin, the diving swallows, the grazing sheep) allowed him not only to map the inscape of the world but also to express his belief in a haptic, corporeal, and vibrant geography. Moreover, they reinforced his conviction that animals are integral to experience; they are not isolated from humanity but rather belong to a hybrid geography.

As the examples of the tiger skin flung over the glacier or the stripes of the zebra's hide visible in the snow signify, wild skins were integral to Hopkins's attempts to represent the inscape of his surroundings. They helped him discover and map the patterns of a prospect or an object. (See Figures 5.5 and 5.6.) As we have seen, sketches accompanying these observations visualize such moments. Animals, though, did not just function as an aid or tool in the mapping of a scene's inscape. They also manifested Hopkins's understanding that the world is best represented through a sensuous geography. Hopkins's use of animal metaphors to map a landscape belongs to his sense of a haptic world in which textures intermingle and in which there is flux or movement. These metaphors contribute to a different kind of map that breaks away from the standard, two-dimensional ones found in travel books, distribution studies, and atlases which freeze the landscape and give no sense of something larger and vibrant.

To begin with, believing that animals are part and parcel of every landscape, Hopkins neither marginalizes nor exiles creatures; instead, he calls on them to help him compose the very essence of a landscape. They are sensuous landmarks, which reflect the complex entanglings of human-animal relations. In this respect, Hopkins almost realizes Lear's desire not only to infuse life into the wild animals he illustrated but also to release these creatures from the private menageries and zoos so that their sensuous bodies can move in time through the spatial map of the earth (see Chapter 3). Unlike the creatures depicted on the period's thematic maps showing the distribution of animals throughout a region or the world, the animals in Hopkins's writings are not regarded as "natural objects to be studied in isolation from their human neighbours" (Philo and Wilbert 4). Rather, the textures and movements of creatures fashion the feel and shape of the world. When, for instance, Hopkins looks at a beech leaf and speaks of its "fur" or points to the "furred branchy tops of the elms" (*Journals* 57), he is not merely reminding himself of its visual appearance but also eliciting the metaphor of an animal's pelt to capture the texture of the trees. And when he compares the look of the glacier to a tiger's skin or the patterns on a snowscape to a zebra's hide, it is not just the visual experience at work; the haptic is also present. The eye gazing at the leaf or the glacier touches what it sees.

It is not surprising then that, when Hopkins describes/maps the view of a town close to the Cumnor hills (quoted in n. 42), he necessarily draws attention not only to the movement but also to the texture of the "fleshly" bodied swallows flying under the bridge. These swallows, "blue and purple above and shewing the amber-tinged breasts reflected in the water," compose the very feel and dimensions of the landscape (*Journals* 134). Their defining shooting movements map the prospect.

These swallows are but one of several examples in which not only the movement of a creature (the rising, the blowing, the running) but also its texture (the slick slipperiness of the fish or the deep softness of a fleece) creates the map:[48] waves rise like fishes' backs (184); a blowing ram's fleece shapes the clouds (212), and a flock of sheep running over the downs determines the land's defining lines as well as the sensation of the land's surface. In an entry written on August 20, 1868, Hopkins describes the moment when he stood in the landscape and watched "the phenomenon of the sheepflock" running over the downs. As an integral part of the landscape, the sheep become one with the running water. Their motions blend and touch to map the scene:

> I saw the phenomenon of the sheepflock on the downs again from Croham Hurst. It ran like the water-packets on a leaf—that collectively, but a number of globules so filmed over that they would not flush together in the exacter comparison: at a gap in the hedge they were huddled and shaking open as they passed outwards they behaved as the drops would do (or a handful of shot) in reaching the brow of a rising and running over. (187)[49]

Without a doubt, animals play an important role in Hopkins's cartographic imagination. They complement and reinforce his understanding that the body mediates the map and that perception is corporeal. Throughout his diaries and journals, Hopkins alters the conventions of zoogeography so that he might demonstrate that these creatures are not an isolated phenomenon segregated from people (as in museums, zoos, and thematic maps that isolate them from all else), but share spaces with humanity and are integral to the larger world. In Hopkins's landscape, animals intertwine with everything else in it (the branches are furry, the glaciers are like tiger skins). Hopkins would have approved of what geographers now refer to as hybrid geography, which acknowledges the fact that animals are integral to the larger world and which desires to discover a cartography able to map this intermingling. He corrects a fault in conventional mapping that neglects to recognize an interface between people and the natural world. By recognizing

---

[48] Phillips, in her study of Hopkins and the Victorian visual world, discusses John Ruskin's influence on Hopkins and calls attention to Ruskin's insistence that if one is to learn the "vital truth," a person must respect motion and draw "the animal in its motion." Only the "dunce thinks [that all things] are standing still" (53).

[49] The "law" is embedded within the motion (whether it be the movement of the flock of sheep or that of water rising and running down). As Hopkins affirms in "God's Grandeur," even though "Toil" has smudged the map, "for all this nature is never spent" (l. 9).

this inter-corporeal bond, he releases his maps from a binary fix so visible in colonial mapping intent on fixing and representing the metropole's power.

The metaphor of animals and their skins also helps him to release the stationary conventional map into a space which is in flux and is experienced from the point of view of one who walks through, observes, and touches its unfolding landscape. He allies the orientation of thinking through space to that of thinking through the body. Hopkins recognizes that geographies should be what Sarah Whatmore has recently termed "fluid, not flat" (3). As the above discussion points out, Hopkins would not have been satisfied with the static nature of the distribution maps in which animals and plants are frozen in time and space. He was attracted, instead, to the map which acknowledges the defining presence of the sensuous body that feels, touches, and moves through a landscape so as to grasp the lines marking the earth's surface/skin. Wanting to chart the vibrant character of a landscape, he occasionally turned to images of animals. Their shapes, their motions, their flight, their feel, their texture, their feathers, and their fur helped make it possible for him to create another kind of map—one which represented the fleshly, living dimension of a landscape and animated the world. In turning his back on conventional maps, which reduced people and places to a flattened, two-dimensional perspective removed from the vagaries of time and feelings, Hopkins broke with the colonial maps (the ones accompanying books about hunting and travel in such places as India and Africa) that seem to have attracted Hardy. He is not drawn to those maps that called upon the presence and hunting of wild animals and their skins to articulate the imperialist's control over the landscape. Hopkins's reference to the tiger's skin to map the glacier is a far cry from the use of wild animals and their collected hides in colonial maps, which, according to Steve Pile and Nigel Thrift, reflected "masculinist and colonial desires to control the land and place its subject within places which it controlled" (Pile and Thrift 5). Hopkins does not obviously use animals as emblems of his or his culture's authority; for him they belong to the breathing, moving earth; they are part of its sensuous geography. They represent a different sort of portrait of the world than that offered at the Belle Vue Zoo, Manchester (Chapter 1) where the animals and their skins are representatives of industrial and imperialistic ventures, and where these captured creatures are segregated and set apart from their surroundings.

Moreover, Hopkins's descriptions of landscape never assume, as did the colonials', that an area has never before been mapped. On the contrary, Hopkins believes that wherever he goes the map is already there (its bones and skin are either under or on the surface as is the landscape's "inscape"); it only needs to be discovered. Hopkins's desire is to excavate the "law" and let the "instress" of it impress itself upon his senses.[50]

---

50 It also has to be recognized, though I do not discuss it in this study, that Hopkins's belief as a Jesuit priest encouraged him to map his landscape by acknowledging the presence of the living, sensuous body of Christ inhabiting the landscape. Hopkins's maps are also

Conclusion

Hopkins's cartographic imagination is simply not tied to imperialist notions of mastery. As a result, consciously, and almost daily, he created maps that display a vibrant interface among animals, humans, and the details of a landscape. His cartography is neither still nor lifeless but proffers, instead, an animistic reading of nature and the world. The landscape itself shifts and stirs. Using moving and fleshly images of animals and skins, Hopkins, in effect, whether he meant to or not, transforms the static world of zoogeography to create a significantly different kind of hybrid map in which animals and humans interact to help form the shape of their surroundings. In his journals at least, no longer are exotic beings simply hunted and displayed as trophies or as scientific specimens. He releases animals from the colonial context of places like the Belle Vue Zoo or museums of natural history. He allows for the experience of reaching, either metaphorically or literally, through the bars or into a collection's glass cases so that one can touch what before had been enclosed. In his maps, Hopkins actually removes these barriers so that animals are no longer banished to a world of exotic otherness but rather are integrated into the landscape of everyday lives.

indebted to his understanding of "God's Grandeur" and a world "charged with the grandeur of God" ("God's Grandeur" l. 1).

Bibliography

Ahmed, Sara and Jackie Stacey, eds. *Thinking Through Skin*. London: Routledge, 2001.

Alberti, Samuel J.M.M., ed. *The Afterlives of Animals: A Museum Menagerie*. Charlottesville: The University of Virginia Press, 2011.

Altick, Richard. *The Shows of London*. Cambridge, MA: Harvard UP, 1978.

"Animal Painting at the Royal Academy." *The Art Journal* (August 1882): 254–6.

Anzieu, Didier. *The Skin Ego*. New Haven: Yale UP, 1989.

Ash, Russell. *Sir Lawrence Alma-Tadema*. London: Pavilion Books, 1989.

"Aptenodytes Pennantis, Esq." *Punch* 48 (1865): 181.

"At the Zoo." *Punch* 53 (1867): 50.

Bain, Alexander. *The Senses and the Intellect*. London: John W. Parker and Sons, 1855.

Ballantyne, R.M. *The Gorilla Hunter: A Tale of the Wilds of Africa*. London: T. Nelson and Sons, 1876.

Barber, Lyn. *The Heyday of Natural History 1820–1870*. Garden City, NY: Doubleday & Company, 1980.

Barlow, Paul. "Facing the Past and Present: The National Portrait Gallery and the Search for 'Authentic' Portraiture." *Portraiture: Facing the Subject*. Ed. Joanna Woodall. Manchester: Manchester UP, 1997. 219–38.

Barnum, P.T. *Lion Jack: A Story of Wild Men and the Capturing of Wild Beasts Showing How Menageries are Made*. London: Sampson Low, Marston, Searle & Rivington, 1877.

Barringer, Tim and Tom Flynn, eds. *Colonialism and the Object: Empire, Material Culture and the Museum*. London: Routledge, 1998.

Bartlett, A.D. *Bartlett's Life Among Wild Beasts in the "Zoo": Being a Continuation of Wild Animals in Captivity*. Comp. and ed. Edward Bartlett. London: Chapman and Hall, 1900.

———. *Wild Animals in Captivity: Being an Account of the Habits, Food, Management and Treatment of the Beasts and Birds at the "Zoo" with Reminiscences and Anecdotes*. London: Chapman and Hall, 1899.

Bayfield, Samuel J. *The Skin, in Health and Disease: Being a Treatise Concerning the Nature of those Diseases Most Frequently Met in Private Practice, with Treatment and Cases*. London: Henry Renshaw, 1865.

Bell, Charles. *The Anatomy and Philosophy of Expressions*. London: John Murray, 1806.

———. *Essays on the Anatomy of Expression in Painting*. London: Longman, Hurst, Reed, and Orme, 1806.

The Belle Vue Collection. Chetham's Library, Manchester.

Benjamin, Walter. "Unpacking My Library." *Illuminations*. Ed. Hannah Arendt. Trans. Harry Zohn. New York: Schocken Books, 1969. 59–68.

Benthien, Claudia. *Skin: On the Cultural Border between Self and the World*. Trans. Thomas Dunlap. New York: Columbia UP, 2002.

Berger, John. *About Looking*. New York: Vintage Books, 1980.

Bernstein, Susan D. "Ape Anxiety." *Journal of Victorian Culture* 6.2 (Autumn 2001): 250–71.

———. "Designs after Nature: Evolutionary Fashions, Animals, and Gender." *Victorian Animal Dreams: Representations of Animals in Victorian Literature and Culture*. Ed. Deborah Deneholz Morse and Martin Danahay. Aldershot: Ashgate, 2007. 65–79.

Betts's Improved Educational Maps: Africa. 1856. [Maps Room, British Library].

Bird's Eye View of the Zoological Gardens, Regent Park. London: H.C. Clarke & Co., 1854.

"A Bloodless Battue." *Punch* 85 (1880): 145.

Blunt, Wilfrid. *The Ark in the Park: The Zoo in the Nineteenth Century*. London: Hamish Hamilton, 1976.

Bolton, Gambier. "Pictures on Human Skin." *The Strand Magazine: An Illustrated Monthly* 5.13 (January–June 1897): 425–34.

Bompas, George C. *Life of Frank Buckland*. London: Smith, Elder, & Co., 1885.

Bouillon, Claude. *Skin: A Living Envelope*. Trans. Alexander Keens. Paris: Découvertes Gallimard, 2002.

Brilliant, Richard. *Portraiture*. London: Reaktion Books Ltd., 1991.

Brown, Captain Thomas. *The Taxidermist's Manual or the Art of Collecting, Preparing and Preserving Objects of Natural History for the Use of Travellers, Conservators of Museums, and Private Collectors*. Glasgow: Archibald Fullarton & Co., 1833.

Browne, Janet. "Biogeography and Empire." *Cultures of Natural History*. Ed. N. Jardine, J.A. Secord and E. C. Spary. Cambridge: Cambridge UP, 1996. 305–21.

Browne, Montagu. *Artistic & Scientific Taxidermy and Modelling*. London: Adam and Charles Black, 1896.

———. *Practical Taxidermy: A Manual of Instruction to the Amateur in Collecting, Preserving, and Setting Up Natural History Specimens of all Kinds*. London: "The Bazaar," [1876].

Browne, Sir Thomas. *Notes for a Lecture on the Skin*. Ed. Geoffrey Keynes. London: John Murray, 1924.

Buckland, Francis [Frank] T. *Curiosities of Natural History*. London: Richard Bentley, 1857.

Burnet, John. *Practical Hints on Portrait Painting*. London: David Bogue, 1850.

Camerini, Jane. "Remains of the Day: Early Victorians in the Field." *Victorian Science in Context*. Ed. Bernard Lightman. Chicago: The University of Chicago Press, 1997. 354–77.

Canton, Gustavus. *The Zoological Garden: An Invitation to the Youth of Both Sexes to Contemplate Animated Nature*. London: Joseph Myers & Co., 1859.

Caplan, Jane, ed. *Written on the Body: The Tattoo in European and American History*. London: Reaktion Books, 2000.

Capuano, Peter J. "Handling the Perceptual Politics of Identity in *Great Expectations*." *Dickens Quarterly* 27.3 (2010): 185–208.

Catlin, George. *Adventures of the Ojibbeway and Ioway Indians in England, France, and Belgium; Being Notes of Eight Years' Travels and Residence in Europe with his North American Indian Collection*. 2 vols. 3rd ed. London: Published by the author, 1852.

Cheetham, Arthur. *Character Reading Practically Explained*. Manchester: Wm Harris & Sons, [1893].

Cheng, Anne Anlin. *Second Skin: Josephine Baker and the Modern Surface*. Oxford: Oxford UP, 2011.

Church, Sally K. "The Giraffe of Bengal: A Medieval Encounter in Ming China." *Medieval History Journal* 7 (April 2004): 1–36.

Classen, Constance, ed. *The Book of Touch*. Oxford: Berg, 2005.

———. *The Deepest Sense: A Cultural History of Touch*. Urbana: University of Illinois Press, 2012.

Cohen, William A. "Arborealities: The Tactile Ecology of Hardy's *Woodlanders*." Victorian Tactile Imagination. Birkbeck College, University of London. July 19–20, 2013. Plenary Lecture.

———. *Embodied: Victorian Literature and the Senses*. Minneapolis: University of Minnesota Press, 2009.

Colley, Ann C. "Edward Lear's Anti-Colonial Bestiary." *Victorian Poetry* 30.1 (Summer 1992): 109–20.

———. "Edward Lear's Limericks and the Reversals of Nonsense." *Victorian Poetry* 26.3 (Autumn 1988): 285–300.

———. "Gerard Manley Hopkins and the Idea of Mapping." *The Search for Synthesis in Literature and Art: The Paradox of Space*. Athens: The University of Georgia Press, 1990. 67–100.

———. "Gerard Manley Hopkins and the Idea of Mapping." *Word & Image: A Journal of Verbal/Visual Enquiry* 4.2 (April–June 1988): 523–8.

———. *Victorians in the Mountains: Sinking the Sublime*. Farnham, Surrey: Ashgate, 2010.

"Coming to the Point." *Punch* 53 (1867): 152.

Connor, Stephen. *The Book of Skin*. London: Reaktion Books, 2004.

Coombe, Annie. *Reinventing Africa: Museums, Material Culture and Popular Imagination*. New Haven: Yale UP, 1994.

Cowling, Mary. *The Artist as Anthropologist: The Representation of Type and Character in Victorian Art*. Cambridge: Cambridge UP, 1989.

Cronin, Jill and Frank Rhodes. *Images of England: Belle Vue*. Stroud, Gloucestershire: Tempus, 1999.

Cross, Edward. *Companion to the Royal Menageries, Exeter 'Change, Containing Concise Descriptions, Scientific & Interesting for the Curious Foreign Animals now in that Eminent Collection Derived from Actual Observation*. London: Tyler and Honeyman, 1820.

Cumming, Roualeyn George Gordon. *Five Years' Adventure in the Far Interior of South Africa.* London: John Murray, 1856.

———. *Five Years of a Hunter's Life in the Far Interior of South Africa with notices of the native tribes, and anecdotes of the chase of the lion, Elephant, Hippopotamus, Giraffe, Rhinoceros, etc.* 2 vols. London: John Murray, 1851.

Darlington, Philip J., Jr. *Zoogeography: The Geographical Distribution of Animals*. Cambridge, MA: Harvard UP, 1957.

Darrah, Henry Zouch. *Sport in the Highlands of Kashmir*. London: Rowland Ward, 1898.

Das, Santanu. *Touch and Intimacy in First World War Literature*. Cambridge: Cambridge UP, 2005.

The Derby Collection, Knowsley Hall.

Dickens, Charles. *Hard Times for These Times*. Ed. Kate Flint. London: Penguin Classics, 2003.

———. *Our Mutual Friend*. Ed. Michael Cotsell. Oxford: Oxford World Classics, 2008.

Dickinson's Comprehensive Pictures of the Great Exhibition of 1851: from the Originals Painted for H.R.H. Prince Albert. London: Dickinson Brothers, 1854.

Diderot, Denis. *An Essay on Blindness, In a Letter to a Person of Distinction*. 3rd ed. London: Printed for J. Baker, [n.d.].

"Discomfiture of old Mr. J—N—S." *Punch* 30 (1856): 144.

Du Chaillu, Paul B. *Explorations and Adventures iin Equatorial Africa*. London: John Murray, 1861.

Fagge, C. Hilton. *Catalogue of the Models of Diseases of the Skin in the Museum of Guy's Hospital*. London: J & A Churchill, 1876.

Field, Tiffany. *Touch*. Cambridge, MA: The MIT Press, 2003.

Fisher, Clemency, ed. *A Passion for Natural History: The Life and Legacy of the 13th Earl of Derby*. National Museums & Galleries on Merseyside. Liverpool: The Bluecoat Press, 2002.

Flanagan, Mary and Austin Booth, eds. *re:skin*. Cambridge, MA: MIT Press, 2006.

Fraser, Louis. Letter to Lord Derby. March 17, 1842. Earl of Derby's Letterbooks, V. 3. National Museums Liverpool.

Freeland, Cynthia. *Portraits and Persons: A Philosophical Inquiry*. Oxford: Oxford UP, 2010.

Frost, Christopher. *Victorian Taxidermy: Its History and Finest Exponents*. Long Melford: Published for the author, 1981.

Gardner, J. *Bird, Animal, and Fish: Stuffing and Preserving: A Manual of Taxidermy for Amateurs*. London: Darton & Hodge, 1866.

Gaskell, Elizabeth. *Mary Barton: A Tale of Manchester Life*. Ed. Macdonald Daly. London: Penguin Classics, 2003.

George, Wilma. *Animals and Maps*. Berkeley: University of California Press, 1969.

Gleanings from the Menagerie and Aviary at Knowsley Hall. [Printed for Private Distribution]. Knowsley, 1846.

Gleanings from the Menagerie and Aviary at Knowsley Hall. [Printed for Private Distribution]. Knowsley, 1850.

Gosse, Edmund W. "Lawrence Alma-Tadema, R.A." *Illustrated Biographies of Modern Artists*. Ed. F.G. Dumas. London: Chapman and Hall, 1882.

Hardy, Thomas. *Tess of the D'Urbervilles*. Ed. Tim Dolan. London: Penguin Books, 2003.

Harvard Library Bulletin (Summer–Fall 2011): 113.

Harvard University Library Open Collections Program. "Contagion: Historical Views of Disease and Epidemics." 2002. http://ocp.hul.harvard.edu/contagion/concepts.hml.

Hauser, Jens, ed. *Sk-interfaces: Exploding Borders—Creating Membranes in Art, Technology and Society*. Liverpool: Fact & Liverpool UP, 2008.

Hazlitt, William. "On *Gusto*." *Selected Writings*. Ed. Ronald Blythe. Middlesex: Penguin English Library, 1982. 201–5.

———. "On the Elgin Marbles." *The Collected Works of William Hazlitt*. Ed. A.R. Waller, Arnold Glover, and William Ernest Henley. London: J.M. Dent, 1902. 326–47.

Hillier, Thomas. *A Hand-Book of Skin Diseases for Students and Practitioners*. London: Walters and Maberly, 1865.

Hoage, R.J. and William A. Deiss, eds. *New Worlds, New Animals: From Menagerie to Zoological Park in the Nineteenth Century*. Baltimore: The Johns Hopkins UP, 1996.

Hopkins, Gerard Manley. *The Journals and Papers of Gerard Manley Hopkins*. Ed. Humphrey House. Completed by Graham Storey. London: Oxford UP, 1959.

Hopkins, Manley. "Essex Elephant." *Once a Week: An Illustrated Miscellany of Literature, Art, Science & Popular Information* 3 (June–December 1860): 53–5 (July 7, 1860).

Hornby, Elizabeth. Diary/Journal. PH/308. Caird Library, National Maritime Museum. Greenwich.

———. Letters. National Museums Liverpool.

Huish, Robert. *The Wonders of the Animal Kingdom Exhibiting Delineations of the Most Distinguished Wild Animals in the Various Menageries of this Country*. London: Thomas Kelly, 1829.

Hunt, Thomas. *A Guide to the Treatment of Diseases of the Skin: with Suggestions for Their Prevention for the Use of the Student and General Practitioner*. London: T. Richards, 1865.

Hyman, Susan. *Edward Lear's Birds*. New York: William Morrow and Company, Inc., 1980.

Illustrated Guide to the Jungle with Descriptions of Indian Animal Life, which has been designed, arranged, and the animals modelled by Rowland Ward,

F.Z.S., Originator and Constructor of the Jungle at the Indian and Colonial Exhibition, 1886. London: Rowland Ward & Co., 1895.

Illustrated London News. August 2–October 25, 1851.

Jackson, Holbrook, ed. *The Complete Nonsense of Edward Lear*. New York: Dover Publications, 1951.

Jennison, George. "History Written by grandson George Jennison, M.A." (1931). The Belle Vue Collection, Chetham's Library, Manchester.

Johnston, W. and A.K. Johnston. *The Physical Atlas: Distribution of Plants and Animals, etc.* 1848. [Maps Room, British Library. Maps 923 (23)].

Johnston's Geographical HandBooks. Edinburgh: W.J.A.K. Johnston, 1902.

Kargon, Robert H. *Science in Victorian Manchester: Enterprise and Expertise*. New Brunswick: Transaction, 2010 [1977].

Keeble, W. *Africa: Showing the Latest Discoveries*. London: Christian Knowledge Society, 1853. [Maps Room, British Library].

"The Keeper's Nightmare." *Punch's Almanack for 1871* (1871): 59.

Kipling, Rudyard. *How the Rhinoceros Got His Skin*. London: Macmillan, 1973.

Kohlstedt, Sarah Gregory. "Reflections on Zoo History." *New Worlds, New Animals: From Menagerie to Zoological Park in the Nineteenth Century*. Ed. R.J. Hoage and William A. Deiss. Baltimore: The Johns Hopkins UP, 1996. 3–18.

Kramer, Andrew E. "Step Right Up Kids." *New York Times*. http://www.nytimes.com/2013/07/11/world/europe/russian-circuses.

Kuhnert, Wilhelm. *Animal Portraiture: Being Fifty Studies By William Kuhnert*. London: Frederick Warne & Co., 1912.

Larsen, Anne. "Equipment for the Field." *Cultures of Natural History*. Ed. N. Jardine, J.A. Secord and E.C. Spary. Cambridge: Cambridge UP, 1996. 358–77.

Lear, Edward. *The Book of Nonsense*. London: Simpkin, Marshall, Hamilton, Kent & Co., 1915.

———. *More Nonsense, Pictures, Rhymes, Botany, Etc.* London: Robert John Bush, 1872.

Lee, Mrs. R. *Taxidermy: or the Art of Collecting, Preparing, and Mounting Objects of Natural History, for the Use of Museums and Travellers*. London: Longman, Brown, Green, and Longmans, 1843.

Letters and notebooks of the 13th Earl of Derby. National Museums Liverpool.

Letters and notebooks of the 13th Earl of Derby and Elizabeth Hornby. National Museums Liverpool.

Lightman, Bernard V. "Spectacle in Leicester Square: James Wyld's Great Globe, 1851–1861." *Popular Exhibitions, Science, and Showmanship, 1840–1910*. Ed. Joe Kember, John Plunkett and Jill A. Sullivan. London: Pickering and Chatto, 2012. 19–39.

Lippincott, Louise. *Lawrence Alma Tadema: Spring*. Malibu, CA: Getty Museum Studies on Art, 1990.

Livingstone, David. Letter to James Risdon Bennett. December 26, 1845. *Livingstone Online*. http://www.livingstoneonline.ucl.ac.uk.

Mackenzie, John, ed. *Imperialism and the Natural World*. Manchester: Manchester UP, 1990.

Mackenzie, Norman H. "Hopkins and Science." *Gerard Manley Hopkins and Critical Discourse*. Ed. Eugene Hollahan. New York: AMS Press, 1993. 81–95.

Map of the Shooting Districts between Shanghai and Wuhu. Shanghai: H.T. Wade and R.A. deVillard, October 1893. [Maps Room, British Library].

Marks, Laura U. *Touch: Sensuous Theory and Multi-Sensory Media*. Minneapolis: University of Minnesota Press, 2002.

Mary, S., P. Benbow and Bonnie C. Hallman. "Reading the Zoo Map: Cultural Heritage Insights from Popular Cartography." *International Journal of Heritage Studies* 14.1 (January 2008): 30–42.

Mayer, David. "The World on Fire: Pyrodramas at Belle Vue Gardens." *Popular Imperialism and the Military*. Ed. John MacKenzie. Manchester: Manchester UP, 1992. 179–97.

"The Meeting of the Zoological Society, Hanover Square." *Punch* 88 (May 23, 1885): 251.

Mendes, D. and H. Ellis. "Joseph Towne (1806–1879), Master-modeller of Wax." *Journal of Medical Biography* 11.4 (November 2003): 212–17.

Merleau-Ponty, Maurice. *Phenomenology of Perception*. London: Routledge & Kegan Paul, 2003.

———. *The Visible and the Invisible: Followed by Working Notes*. Ed. Claude Lefort. Evanston: Northwestern UP, 1968.

Merriford, Mrs. *Practical Directions for Portrait Painting in Water-Colours*. London: Winsor and Newton, 1851.

Michie, Helena. *Flesh Made Word: Female Figures and Women's Bodies*. New York: Oxford UP, 1987.

Midgley, Mary. *Animals and Why They Matter*. Athens: The University of Georgia Press, 1998.

Mitchell, D.W. *A Popular Guide to the Gardens of the Zoological Society of London*. London: R.S. Francis, 1852.

Mitchell, P. Chalmers. *Centenary History of the Zoological Society of London*. London: Printed for the Society, 1929.

Montagu, Ashley. *Touching: The Human Significance of the Skin*. New York: Harper & Row, 1986.

Morning Post. April 22, 1886.

Morris, P.A. *A History of Taxidermy: Art, Science and Bad Taste*. Ascot, Berkshire: MPM Publishing, 2010.

Murray, Henry. *The Art of Portrait Painting in Oil Colours with Observations on Setting and Painting the Figure*. London: Winsor and Newton, 1851.

NAVSA 2013: Evidence: The North American Victorian Studies Association Annual Conference. Pasadena, California. October 23–6, 2013.

Nicholls, Robert. *The Belle Vue Story*. Manchester: New Richardson, 1992.

Noakes, Vivien, ed. *Edward Lear 1812–1888*. London: Weidenfeld and Nicolson, 1985.

Nott, J. Fortune. *Wild Animals Photographed and Described: Illustrated by Phototype Reproductions of Photographic Negatives Taken from Life*. London: Sampson Low, Marston, Searle & Rivington, 1886.

O'Farrell, Mary Ann. *Telling Complexions: The Nineteenth-Century English Novel and the Blush*. Durham: Duke UP, 1997.

Oswell, W. Edward. *William Cotton Oswell: Hunter and Explorer, the Story of his Life*. 2 vols. New York: Doubleday, Page & Co., 1900.

Owen, Richard. *Memoir on the Gorilla*. London: Taylor and Francis, 1865.

Owen, Richard, Rev. *The Life of Professor Owen*. 2 vols. London: John Murray, 1894.

Palmer, A.H. *The Life of Joseph Wolf Animal Painter*. London: Longmans, Green, & Co., 1895.

Patchett, Merle M. "Putting Animals on Display: Geographies of Taxidermy Practice." Dissertation. University of Glasgow Department of Geographical and Earth Sciences, 2010.

Patterson, Mark. *The Senses of Touch: Haptics, Affects and Technologies*. Oxford: Berg, 2007.

Peck, Robert McCracken. "The Natural History of Edward Lear." *Harvard Library Bulletin* 22.2–3 (Summer–Fall 2011): 1–68.

Perricone, Christopher. "The Aspiration to the Condition of Touch." *Philosophy and Literature* 30.1 (2006): 229–37.

Phillips, Catherine. *Gerard Manley Hopkins and the Victorian Visual World*. Oxford: Oxford UP, 2007.

Philo, Chris and Chris Wilbert, eds. *Animal Spaces, Beastly Places: New Geographies of Human-Animal Relations*. London: Routledge, 2000.

Physical Map of the World from the Discoveries of the Most Eminent Modern Geographers. London: James Reynolds, 1849. [Maps Room, British Library 950 (34)].

Pictorial Museum of Animated Nature. V. 1: Mammals and Birds. London: Charles Knight and Co., 1843.

Picturesque Guide through the Regent's Park with Accurate Descriptions. London: J. Limbird, 1829.

Pile, Steve and Nigel Thrift, eds. *Mapping the Subject: Geographies of Cultural Transformation*. London: Routledge, 1995.

Plan of the Garden of the Zoological Society within the Area of the Regent's Park. London: 1841. [Maps Room, British Library].

Poliquin, Rachel. *The Breathless Zoo: Taxidermy and the Culture of Longing*. University Park: The Pennsylvania State UP, 2012.

A Private Life of an Eastern King by a Member of the Household of his Late Majesty, Nussir-V-Deen, King of Oude. New York: Redfield, 1855.

Progressive Lessons Intended to Elucidate the Art of Portrait Painting in which is Introduced a Systematic Arrangement of the Colours and Tints Used in Flesh, Draperies, and Back-Grounds with Examples. London: Thomas Clay, 1824.

"A Prospecte of Ye Zoological Societye: Its Gardens." *Punch* 17 (1849): 192.

Reader, Simon. "'Among the loose papers': Gerard Manley Hopkins's Idle Documents." The North American Victorian Studies Association Annual Conference. Pasadena, CA. October 23–6, 2013. Conference presentation.

Reeve, Lovell. *Letter to the Right Honourable The Earl of Derby, K.G., D.C.L., on the Management, Character, and Progress of the Zoological Society of London*. London: Reeve, Brothers, 1846.

Reports by the Juries on the Subjects in the Thirty Classes into which the Exhibition was Divided. London: Clowes, 1852.

Ritvo, Harriet. *The Animal Estate: The English and Other Creatures in the Victorian Age*. Cambridge, MA: Harvard UP, 1987.

———. "The Order of Nature: Constructing the Collection of Victorian Zoos." *New Worlds, New Animals: From Menagerie to Zoological Park in the Nineteenth Century*. Ed. R.J. Hoage and William A. Deiss. Baltimore: The Johns Hopkins UP, 1996. 43–50.

———. "Zoological Nomenclature and the Empire of Victorian Science." *Victorian Science in Context*. Ed. Bernard Lightman. Chicago: The University of Chicago Press, 1997. 334–53.

Robbins, Bruce. *The Servant's Hand: English Fiction from Below*. New York: Columbia UP, 1986.

Robinson, Arthur H. *Early Thematic Mapping in the History of Cartography*. Chicago: Chicago UP, 1982.

Rodaway, Paul. *Sensuous Geographies: Body, Senses and Place*. London: Routledge, 1994.

Rowley, John. *The Art of Taxidermy*. New York: D. Appleton and Company, 1898.

Ruskin at Walkley: Reconstructing the St. George's Museum. University of Sheffield, 2011. Web. 2014.

"Sanitary Measures." *Punch* 56 (1869): 140.

Sauer, Gordon C., ed. *John Gould the Bird Man: Correspondence with a Chronology of his Life and Works*. Vol. 1. London: The Natural History Museum, 1998.

Scherren, Henry, ed. *Walks and Talks in the Zoo*. London: The Religious Tract Society, [1901].

———.*The Zoological Society of London*. London: Cassell and Company, [1905].

Schiebinger, Londa. "Gender and Natural History." *Cultures of Natural History*. Ed. N. Jardine, J.A. Secord and E.C. Spary. Cambridge: Cambridge UP, 1996. 163–77.

Schlossman, Beryl. *Objects of Desire: The Madonnas of Modernism*. Ithaca: Cornell UP, 1999.

Sennett, Richard. *The Craftsman*. New Haven: Yale UP, 2008.

Serres, Michel. *Five Senses: A Philosophy of Mingled Bodies*. Trans. Margaret Sankey and Peter Cowley. London: Continuum International Publishing Group, 2008.

"Si Non E Vero." *Punch* 80 (April 2, 1881): 146.

"Specimens not yet included in the Collection at Regent's Park." *Punch* 56 (1869): 258.

Spencer, Robin. "Whistler's 'The White Girl': Painting, Poetry and Meaning." *The Burlington Magazine* 140.1142 (May 1998): 300–311.

Stackhouse, Heather and Daniel Hyams. *Belle Vue's Manchester Playground*. Cheshire: At Heart Publications. Manchester Evening News, 2007.

Stanley, Edward Smith, 13th Earl of Derby. Letters and notebooks of the 13th Earl of Derby. National Museums Liverpool.

Stewart, Susan. *On Longing: Narratives of the Miniature, the Gigantic, the Souvenir, the Collection*. Durham: Duke UP, 1993.

Stoichita, Victor L. *A Short History of the Shadow*. London: Reaktion Books, 1997.

Sussman, Herbert L. *Victorian Technology: Invention, Innovation, and the Rise of the Machine*. Santa Barbara: Praeger, 2009.

———. *Victorians and the Machine: The Literary Responses to Technology*. Cambridge: Harvard UP, 1969.

Swainson, William. *Animals in Menageries*. London: Longman, Orme, Brown, Green, & Longmans, 1838.

———. *Taxidermy, a Bibliography, and Biography*. London: Longman, Orme, Brown, Green, & Longmans, 1840.

Swayne, Major H.G.C. *Seventeen Trips through Somaliland and a Visit to Abyssinia*. 3rd ed. London: Rowland Ward, 1903.

Tallis, Raymond. *The Hand: A Philosophical Inquiry into Human Being*. Edinburgh: Edinburgh UP, 2003.

———. *Michelangelo's Finger: An Exploration of Everyday Transcendence*. London: Atlantic Books, 2010.

Tree, Isabella. *The Ruling Passion of John Gould: A Biography of the Bird Man*. London: Barrie & Jenkins, 1991.

Vevers, Gwynne. *London Zoo: An Anthology to Celebrate 150 Years of the Zoological Society of London*. London: Bodley Head, 1976.

The Victorian Tactile Imagination Conference. Birkbeck College. July 10–20, 2013.

"Visiting a Private Collection Mistakes 'Peter' the Great Horned Owl for a Stuffed Cat." *Punch* 30 (April 12, 1856): 144.

Wallace, Alfred Russell. *The Geographical Distribution of Animals with a Study of the Relations of Living and Extinct Faunas as Elucidating the Past Changes of the Earth's Surface*. New York: Hafner Publishing Co., 1962 [1876].

———. *The Malay Archipelago*. Introd. John Bastin. Singapore: Oxford UP, 1986.

Ward, Priscilla. *Contagious: Cultures, Carriers, and the Outbreak Narrative*. Durham: Duke UP, 2008.

Ward, Rowland. *The Sportsman's Handbook to Practical Collecting, Preserving, and Artistic Setting-Up of Trophies and Specimens to which is added a Synoptical Guide to the Hunting Grounds of the World*. London: Rowland Ward and Co., 1894.

Waterton, Charles. *Wanderings in South America, the North-West of the United Sates and the Antilles in the Years 1812, 1816, 1820, & 1824 with original instructions for the perfect preservations of Birds, etc. for Cabinets of Natural History*. London: B. Fellows, 1828.

Whatmore, Sarah. *Hybrid Geographies: Natures, Cultures, Spaces*. London: Sage, 2002.

What on Earth: A Collection of Funwear & Delightful Diversions. Mail Order Catalogue, 2012.

Whistler, James McNeill. *The Correspondence of James McNeill Whistler*. University of Glasgow. http://www.whistler.art.gla.ac.uk/correspondence/subject/display.

Whitaker, Katie. "The Culture of Curiosity." *Cultures of Natural History*. Ed. N. Jardine, J.A. Secord and E.C. Spary. Cambridge: Cambridge UP, 1996. 75–90.

Wilde, Oscar. *The Picture of Dorian Gray*. Ed. Joseph Bristow. Oxford: Oxford World Classics, 2008.

Wilson, Erasmus. *A Practical Treatise on Healthy Skin: with Rules for the Medical and Domestic Treatment of Cutaneous Diseases*. London: John Churchill, 1844.

Wireless. Mail Order Catalogue, 2013.

Wollstonecraft, Mary. *Letters Written in Sweden, Norway, and Denmark*. Ed. Tone Brekke and Jon Mee. Oxford: Oxford World's Classics, 2009.

Wood, Samuel. *The British Bird Preserver*. London: Frederick Warne & Co., [1877].

Woodall, Joanna, ed. *Portraiture: Facing the Subject*. Manchester: Manchester UP, 1997.

Zaniello, Tom. "Alpine Art and Science: Hopkins's Swiss Adventure." *The Hopkins Quarterly* 27.1/2 (2000): 3–17.

Zimmern, Helen. *Sir Lawrence Alma Tadema, R.A.* London: George Bell & Sons, 1902.

Zoological Keepsake; or Zoology, and the Garden and Museum of the Zoological Society, for the Year 1830. London: Marsh and Miller, [1830].

Zoological Map of the World, Shewing the Geographical Distribution of Animals. London: William Spencer, November 5, 1845. [Maps Room, British Library. Maps 950 (7)].

Zoological Map Showing the Distribution of Animals over the World. London: J. Reynolds, [1849?] [Maps Room, British Library. Maps 950 (33)].

Index

Adamson, Daniel 28
Addison, Thomas 11
adventure 5–7, 12, 56–7, 63, 65, 70, 80, 87, 89, 126, 131, 149, 152, 154, 159, 166
  fiction 5, 12, 56–7, 89
  narrative 56–7, 63, 65, 70, 126
Africa (continent) 49, 53, 56–7, 63, 65, 73, 75, 77, 96, 121, 154–5, 159, 161, 180
African people 12, 32–3, 69, 130
Ahmed, Sara 14, 32, 35, 128–9, 134, 136
Alienness 12, 34, 46, 50, 52, 54, 71, 73, 79, 107, 149, 159
Alma-Tadema, Lawrence (Sir) 5–6, 137–43, 146
  In The Tepidarium (1881) 6, 137–43, 146
Altick, Richard 134
anatomy 12, 18, 33, 82, 94, 106, 155
animal skin. *See also* skin
  and mapping 6, 15, 46 149–81
  and preparing 27, 43, 63, 68–9, 71, 73, 87, 98, 100, 104, 106
Ansdell, Richard 108
Anzieu, Didier 13–14
Arkwright, Richard 29
Ash, Russell 139–40
Asia (continent) 23, 53, 60, 63, 65, 75, 89, 91, 94, 104, 106, 117, 156–8, 161, 180
Audubon, John James 87–8, 107, 109
Australia (continent) 75, 161, 163
authority 4, 28, 47, 50, 54, 65, 67, 69, 74, 76, 90, 119, 127, 130, 180

Baedeker's guides, 21, 166
Bain, Alexander 13, 15, 129
Baker, Josephine 15, 130–32
Ballantyne, R.M. 56–7, 71
Banfield, Marie 129, 165
Barber, Lynn 49, 51–3, 72, 74, 85
Barlow, Paul 30
Barnum, P.T. 56–7, 65
barrier 4,6, 14, 23, 38, 40, 62, 121–34, 147, 149, 181. *See also* cage
Bartlett, A.D. 57, 59, 62, 64, 67–8, 94, 100, 104
Bates, Henry Walter 52
"The Battle of Alma" panorama (1896) 23, 27. *See also* Belle Vue Zoo
"The Battle of the Nile" panorama (1862) 23. *See also* Belle Vue Zoo
Bayfield, Samuel J. 11
  The Skin, in Health and Disease: Being a Treatise Concerning the Nature of those Diseases Most Frequently met with in Private Practice, with Treatment and Cases (1867) 11
Beer, Gillian 165
Bell, Charles (Sir) 12, 39
  Essays on the Anatomy and Expression in Painting (1806) 12
  The Anatomy and Philosophy of Expressions (1824) 12
Bell, Thomas 109
Belle Vue Zoo (Manchester) *xi–xii*, 5, 17–35, 37–48, 62, 96, 127, 149, 180–81
Benjamin, Walter 50, 74, 85
Benthien, Caludia 14, 32–3, 45, 64, 66, 136
Berenson, Bernard 134–5
Berger, John 31–2, 116
Bernstein, Susan D. 35, 40
Berzine, Zane 129
Bethell, Augusta "Gussie" 87
Betts (mapmaker) 154
Bingley, William 130
birds 1, 4–6, 9. 17, 23, 33, 38, 40, 42–3, 47, 49, 51, 53, 56–7, 60–61, 63–5,

67–71, 73–5, 87–8, 90, 93, 95, 97–8, 100, 104–14, 116, 118–19, 150, 154–6, 159–61, 163, 173
auk 161
Bird of Paradise 69
cockatoo 161
condor 83–4, 161, 163
cormorant 106
crane 75, 89, 113
crow 106
cygnet 60, 79, 81–4, 161
eagle 76, 161
finch 163
flamingo 1
guinea fowl 64
harpy eagle 161
hummingbird 57, 65, 69
kiwi 107–8
lyre bird 161
macaw 81, 111–12, 161
ostrich 32, 65, 67, 75–6, 140, 161
owl 1, 54, 67, 102, 104, 111–16, 121
parrot 33, 61, 82, 84, 87–9, 103, 110–13, 116–17, 119, 161
partridge, 79, 82, 84
pelican 38, 113, 161
penguin 35, 41, 79, 82–3
pheasant 73, 153, 156
pigeon 79, 82
plover 84
rhea 160–61
Rio blackbird 84
seagull 106
sparrow 79
stork 113
swallow 173, 178–9
swan 60, 79, 81–4, 161
toucan 161
vulture 57, 156, 161
wren 169
Blackfriars (London) 11
Blake, William 10, 71
blush 95, 176–7. *See also* complexion; skin
Boer War 23, 28
Bolton, Gambier 13
"The Bombardment of Algiers" panorama (1852) 23. *See also* Belle Vue Zoo
Bonaparte, Napoleon 42–3
Bond, Edward 166
bones 43–5, 101, 155, 166–6, 176
Bonnard, Pierre 137
Booth, Austin 14, 32–5
Bouillon, Claude 14, 32, 34–5, 128, 132, 134
Bowdich, Thomas Edward 63
breeding 75, 88, 116
brickworks 21–2
Brilliant, Richard 31
Bristol Zoo 18
British Library *xi–xii*, 153–4, 162–4
British Museum 49, 57, 72–3, 112, 134
Brooke, James (Sir) 60
Brown, Thomas (Captain) 44, 70–73, 98–9
Browne, Janet 52,
Browne, Montagu 56, 68, 104,
Browne, Thomas (Sir) 10
Notes for a Lecture on the Skin (1676) 10
Buckland, Frank 61, 104–07
Bullock Museum (Liverpool) 77
Burma 171–2
Burnet, John 39–40, 135, 146
Practical Hints on Portrait Painting (1850) 39, 145

cage 6, 23, 31–3, 38, 41, 43, 49, 61–2, 84, 94, 101, 111, 121, 126–8, 159. *See also* barrier
Caird Library *xi*, 79
Cambridge University *xi–xii*, 101, 116
Camerini, Jane 51–2
Canton, Gustavus 39
The Zoological Garden: An Invitation to the Youth of Both Sexes to Contemplate Animated Nature (1859) 39
Caplan, Jane 13
Capuano, Peter J. 136
caress 6, 121, 123, 126, 132, 136–8, 140, 145–7. *See also* touch
Carlyle, Thomas 30
cartography 159–81
Catlin, George 65, 126, 131–2
Central America 75, 80, 163
character 12–13, 33, 40–41, 48, 64, 94–6, 106, 120, 136, 151, 161, 166, 180

chartists 21
Cheetham, Arthur 12
Cheng, Anne Anlin 15, 130
Chethem's Library (Manchester) *xi–xii*, 19, 23, 25–6, 38, 43–4, 48, 127
Chile 80, 84
China 60, 75, 156
Christian Knowledge Society 154
Church, Sally K. 60
Circus 20, 56, 121, 131
  Nikulin Circus (Moscow) 121
  Ringling Brothers and Barnum and Bailey 56
Clarke, H.C. of H.C. Clarke and Co. (map publishers) 159
Classen, Constance 16, 130, 134
classification 4, 6, 11–13, 41, 49, 50–54, 62, 74–5, 85, 93, 106, 113–14, 119–20, 132, 150, 159. *See also* Cuvier, Georges
cloth and clothing 4, 6, 15, 32, 64–5, 83, 95, 126, 131, 138, 142–3, 145, 153, 176
  versus hair/skin 4, 32, 64–5, 131, 143
Cohen, William A. 96, 129, 136
collecting and collections 3–6, 9, 12, 14, 16–18, 21–3, 27–9, 32, 40, 42–3, 46, 49–53, 56–7, 60, 62–4, 66–79, 81–5, 87–9, 91, 93–4, 98, 102, 106–11, 113–14, 116–17, 120, 133–4, 149, 155–7, 161, 163, 165, 179–81
Colonial and Indian Exhibition of 1886 91, 94, 104, 106
colonial times 4–7, 23, 28, 33, 50, 52–3, 63, 69, 74, 78, 80, 85, 89, 90–91, 93–4, 104, 106, 113, 116–17, 120, 128, 149, 152, 154–5, 159, 163, 166, 180–81
Colonial adventurers' maps 6–7, 149, 152, 155, 180
commodification of skin 6, 91, 93, 113, 116, 120
complexion 13, 16, 95, 149, 151–2, 154–5, 161, 177
Connor, Steven 14, 32–3, 35, 44, 65–6, 97, 102, 128, 130, 135–6, 147
contagion 27, 50, 70–73, 132
contamination 6, 67, 73
Coombe, Annie 159
cosmetics 13, 136
Cox, Kimberly *xi*, 136
Crimean War 23, 28, 46
Cronin, Jill 22
Cross, Edward 32
  Companion to the Royal Menagerie (1820) 32
Crystal Palace 67
cultural anxiety 137, 145–6,
cultural knowledge 94, 110, 131, 150, 160, 170
Cumming, Roualeyn George Gordon 56, 63–5, 73, 96, 155, 157
Currie, Walter T. 49
Cuvier, Georges 41

d'Agrate, Marco 10
Danson, George 19–20, 24
  Manchester from Belle Vue (1861) 19–20, 24
Darlington, Jr., Philip J. 160
Darrah, Henry Zouch 157–8
Darwin, Charles 34, 51–2, 72, 80–81, 137, 160, 163
Darwin, Erasmus 128
Das, Santanu 15, 129
David, Gerard 10
  The Flaying of Sisamnes (1498) 10
David, Jacques-Louis 30
de Villard, R.A. (map publisher) 156
"The Defense of Lucknow" panorama (1863) 23. *See also* Belle Vue Zoo
"Delhi: Storming the Kashmir Gate" panorama (1906) 24–5. *See also* Belle Vue Zoo
dermatology 10–11. *See also* skin
Deutscher, Penelope 128
Diamond, Sara 34
Dibb, Robert 23
Dickens, Charles 5, 19–21, 93–8, 136
  Hard Times 19–21
  Oliver Twist 95
  Our Mutual Friend 93–8
Diderot, Denis 15, 133
display 1, 3–6, 9–12, 14, 17–25, 27–34, 40, 42–3, 45, 47–9, 51, 56–7, 62,

64–5, 67, 70, 74, 78, 82, 85, 89, 94–6, 100–101, 103–06, 109–11, 114, 121, 126, 128, 130–31, 134, 136, 140, 142, 149, 152, 154, 157, 159–60, 165, 170, 176, 181
of fireworks 21, 23–7, 47
of skin 1, 5–6, 9–10, 12, 14, 17, 31, 33–4, 42, 45, 49, 65, 67, 70, 103, 111, 114, 130, 140, 149
Disraeli, Benjamin 30
distribution of animals 6, 149, 160–65, 169, 178, 180. *See also* zoogeography
Du Chaillu, Paul B. 56–7, 63, 69–70, 104
Duke of Wellington 28–9, 31

Edwards, John C. *xi*, 109
Egerton, Philip (Sir) 28, 72
elephantitis 11
Ellis, H. 11
emblem 4, 20, 23, 31, 33, 40, 120, 171, 180
Empire *xii*, 4–5, 12, 17–50, 52, 62, 64, 74, 128, 149, 154, 159
encounter (site of) 1, 6, 9, 12–14, 16, 31, 35, 73, 123, 126, 129, 132, 136, 145–6, 154, 176
England 7, 10, 20, 22, 60–61, 76, 80, 83–4, 87, 89, 93, 100, 107, 120, 126, 131, 136, 157
envelope (of the body) 4, 9, 13–14, 34–5, 39, 40, 43–6, 49, 66–7, 71, 85, 96–7, 101, 116, 130, 134, 140, 143, 151, 157, 161, 176–7
eroticism 137–40, 142–3, 145
Europe (continent) 17, 24, 53, 110, 113, 126, 130–31, 171
Everest, Sophie 48
evolution 6, 130, 137, 160
exhibitions 3, 19, 30, 47, 56, 65, 91, 94, 104–06, 131, 159
exoticism 1, 3–7, 9 14, 17, 19–20, 22–3, 31, 33–4, 49–50, 52–4, 73–4, 89–91, 97, 107, 117, 120–21, 123, 127–30, 137, 145, 147, 152, 154–5, 159–60, 181

Fagge, C. Hilton 11–12. *See also* Gordon Museum of Guy's Hospital
"The Fall of Sebastopol" panorama (1855) 23. *See also* Belle Vue Zoo
fascination 1, 6, 19, 50, 75, 77, 80, 93, 149
feather 1, 4, 6, 9, 17, 31, 33, 38, 44, 49,68, 70–71, 73, 79, 97–8, 103–04, 106–07, 110–12, 114, 131, 133, 139–40, 169–70, 180
Field, Tiffany 129
fingerprinting and identity 13
firework displays 21, 23–5, 27, 47
Fisher, Clemency *xi*, 75, 78, 82, 88, 108, 111, 113
Fitzroy Square (London) 11
Flanagan, Mary 14, 33–5
flaying 10, 57, 73, 89, 92, 140, 146. *See also* skin
Foucher, Léon 41
Fourment, Hélène 130, 132, 140
Frankenstein 98
Fraser, Hilary 134
Fraser, Louis 69, 78
Freeland, Cynthia 28, 112
Fry, John (Reverend) 75–7

Gabon 57, 64
Galton, Francis 13, 155
Gardner, J. 79, 90, 97, 100
Gaskell, Elizabeth 9, 20, 41, 51
Mary Barton 20, 41, 51
Gates, Barbara T. 52
gaze 5, 7, 9,13, 17, 23–4, 32, 32–5, 38, 40, 42, 74, 78, 85, 93–4, 96, 101–02, 112–13, 116, 133, 137, 142, 167, 178
geography 6, 52, 149–50, 155, 159–61, 163, 165–6, 175, 178–88. *See also* zoogeography
sensuous geography 150, 165, 175, 177–8, 180
George, Wilma 159–60
Gilbert, John 63, 107
Gilbert, Pamela K. 129
Giordano, Luca 10
Apollo and Marsyas (1650) 10
glass eye 1, 101
Gordon Museum (Guy's Hospital, London) *xi*, 11
Gosse, Edmund 142

Gosse, Ellen 142
Gosse, Philip 49, 74, 85
Gould, John 49, 56, 63–5, 69, 72, 87, 93, 107, 109–10, 113, 119–20
Gray, John Edward 53, 108, 112, 166
Great Exhibition (1851) 3, 30, 56, 65, 105
    Reports by the Juries 3
grotto 21, 23
"gusto" 142, 145

Hallam, Arthur Henry 136
hand 6, 12–13, 15–16, 21, 39–40, 68, 73, 98, 101, 104, 108, 121–3, 126, 128–9, 131–40, 143, 145–47, 152, 155, 176–7
    and touch 13, 15–16, 73, 121–3, 128–9, 131–40, 177
haptic sight (haptic visuality) 6, 15–16, 133–5, 137, 177–8
Hardy, Thomas 5–7, 149–54, 176–7, 180
    Tess of the D'Urbervilles 149–54, 177
Hauser, Jens 14, 129
Hawkins, Benjamin Waterhouse 108, 113
Hazlitt, William 121, 142, 145, 147
    "gusto" 142, 145, 147
hide 1, 3–4, 6–7, 9, 32–4, 40, 45, 48–9, 56–7, 63, 65, 71, 91, 95–6, 101, 103–04, 110, 121, 129–30, 132, 137, 139, 145–6, 149–50, 152–4, 156–7, 161, 166–7, 178, 180
Higgins, Lesley 167
Hilliar, Thomas 11
    Hand-Book of Skin Diseases for Students and Practitioners (1865) 11
Hopkins, Gerard Manley 5, 7, 150, 165–81
    "God's Grandeur" 174, 179, 181
    inscape 168, 170–71, 175–6, 178, 180
    "Pied Beauty" 174
horn 53, 170
Hornby, Elizabeth *xi*, 5, 50, 74, 78–85, 87–8, 93, 98, 100, 104
Hornby, Phipps (Vice Admiral) 80, 87–8
Horniman Museum (London) 106
Hospital for Diseases of the Skin (Blackfriars, London) 11
Houghton Library (Harvard University) *xii*, 117–18
Hubbard, James Henry 106
Hunt, Thomas 11
    A Guide to the Treatment of Diseases of the Skin: with Suggestions for Their Prevention for the Use of the Student and General Practitioner (1865) 11
hunting 5–6, 11, 34, 51, 56–7, 63–5, 71, 73, 83, 90, 96, 98, 106, 116, 149, 152–7, 159, 161, 180–81
hybrids 50, 75, 77, 130, 146, 178–9, 181
Hyman, Susan 111

identity 1, 4–5, 9–10, 12–14, 16–18, 28, 32, 45–6, 50, 64, 74, 94–6, 108, 116, 130, 136
Illustrated London News (newspaper) 29, 104
imagination 7, 13, 16, 96, 129, 134–4, 138, 149, 154, 165, 171, 173–6, 179, 181
    cartographic imagination 7, 149, 154, 165, 171, 173–6, 179, 181
    tactile imagination 16, 129, 134–5
imperialism 17–18, 23–4, 28, 33, 42, 45, 47, 49–53, 56–7, 62, 70, 116, 119–20, 154, 180–81
India 23, 53, 63, 65, 75, 89, 91, 94, 104, 106, 117, 157–8, 161, 180
Indians (American and South American) 6, 42, 80, 126, 131–2, 147, 177
inscape 168, 170–71, 175–6, 178, 180
insects 42, 51, 70–73, 78, 81–2, 156

Jackson, Christine E. *xi*, 78, 87, 89, 108
Jardine, William (Sir) 87, 109
Jennison, George 18, 20, 24, 27, 40, 41–3, 46, 48
Jennison, John 20–25, 27, 46, 48
Johnston, W. and A.K 161
    The Physical Atlas 161
Jugend magazine 53–4

Kargon, Robert H. 41
Keeble, W. (mapmaker) 154
"King Oude's Fighting Tiger" 42–3

Kingsley, Mary 57
Kipling, Rudyard 5, 95
Knowsley Hall *xi*, 69, 75–8, 82–4, 87–9, 93, 108–13, 120
Knowsley Menagerie Museum 88, 111, 113
Kohlstedt, Sarah Gregory 49, 52
Kramer, Andrew E. 121
Kuhnert, Wilhelm 65

labeling 4, 24, 47, 53, 56, 62–3, 74, 78, 94, 96, 159, 170
Lancashire 21, 41, 47
  Lancashire mills 21
landmark 154–6, 159, 166, 173, 175, 178
landscape 6–7, 15, 18, 20, 46, 52–3, 88–9, 119, 133, 149–51, 153–5, 157, 159, 161, 163, 165–7, 169–71, 173–81
Largen, Malcolm 77
Larsen, Anne 62
Lear, Edward 5–6, 53, 87–93, 107–20, 178
  Book of Nonsense (1846) 88
  "Manypeeplia Upsidownia" 53
  "Old Person of Crowle" 113
  "Portraites of the inditchenous beestes of New Olland" 119
  "The Bay-headed Parrot" 112
  "The Black Stork" 113
  "The Dark Blue Bird" 119
  "The Light Green Bird" 119
  "The Pelican Chorus" 113
  "The Pink Bird" 119
  "The Red and Yellow Macaw" 112
  "There was an Old Man of Dumblane" 113
  "There was an Old Man of Dumbree" 114
  "There was an Old Man on the Border" 119
  "There was an Old man with an Owl" 114
  "The Spectacled Owl" 111–12
  "The Story of the Four Little Children Who Went Round the World" 5, 87–8, 90–93, 112
Lee, Mrs. R. 63–4, 68, 79
Leverian Museum (London) 77
Lightman, Bernard V. 23
Lindquist, Jason H. 165
Linnaean Society 75, 87, 163
Lippincott, Louise 142
Livingstone, David 106, 155–6
London Zoological Society Gardens 13, 18, 41, 45, 49, 51, 57, 60–61, 65, 68, 72, 75–6, 78, 87, 91, 96, 109, 114, 126–7, 131, 159

Macdonald's tattoo studio (Jermyn Street, London) 13
MacKenzie, John M. 51–2
Mackenzie, Norman 165, 167
Macleod, Roy 52
magic lantern 27–8, 51
Malaysia (use also Malay) 49, 67–9, 163
mammals 33, 63–4, 77, 81, 88, 90, 93, 95, 111–12, 154, 160–61
  ant eater 161
  antelope 96, 156, 161
  ape 35, 39–40, 113, 116, 123, 161
  armadillo 161
  baboon 22
  bear 6, 46, 61–2, 64, 69, 71, 77, 123, 126, 130–32, 138–40, 146, 161, 166
  beaver 123
  Bengal tiger 33–4, 60, 94–5
  boar 78, 156, 161
  buffalo 65, 75, 108, 161
  camel (dromedary) 25, 27, 32, 57, 60, 72, 93, 123–4, 161
  chimpanzee 22, 39, 62, 94, 121, 123
  chinchilla 84, 161
  Coypu (river rat) 41, 83–4
  deer 53, 60, 76–7, 83–4, 108
  elephants 22, 25, 27, 32–4, 43–4, 47, 53–4, 60, 62, 67–8, 91, 96, 123, 149, 154–5, 157, 161, 165–6
  ermine 158
  fat-tailed dunnot 130
  ficogna 161
  fox 106
  giraffe (or cameleopard) 38, 53–4, 57, 60, 65, 93, 123
  gorilla 49, 56–9, 62, 64, 71, 94, 104
  hare 106, 153
  hart 154
  hedgehog 119

hippopotamus 34, 38, 42–3, 54, 57, 60, 92, 123–4, 132, 155, 161
horse 31, 80–81, 96, 161
hyena (hyaena) 32, 65, 123
hyrax 111
jackal 161
jaguar 32, 161
kangaroo 61, 63, 89, 161
kangaroo rat 161
koodoo (antelope) 156
lemur 33, 123, 161
leopard 32, 77, 123, 126, 139, 153
lions 7, 32, 42–3, 46–7, 49, 53–7, 60–62, 64, 75–7, 89, 106, 121, 123, 129–32, 155–7, 161, 166
llama 76, 160–61
mandrill 48
monkey 1, 23, 32, 35, 42, 45, 47, 53, 61, 69, 104, 126–7, 160–61
oppossum 111, 160
orangutan 23, 29, 60, 161
panther 32, 156
peccary 160
porcupine 32, 35
puma 161
ram 169, 179
rhinocerous 34, 57, 62, 65, 72–3, 89–92, 95, 112, 120–23, 127, 132, 147, 152, 156–7, 161
sheep 38, 76, 84, 131, 170, 175, 178–9
sloth 45, 161
tapir 123, 125
vicuna 76
walrus 106
water buffalo 108, 161
wildcat 1
wombat 97
zebra 77, 101–02, 113, 116, 157, 167–70, 178
Manchester *xi–xii*, 5–6, 17–48, 51, 127, 131–3, 147, 160, 180
and empire *xii*, 5–6, 17–48
and industry/commerce *xii*, 17–48
Manchester Mechanics' Institute 41
manikin 101–02
manufacturing 4, 17, 19–22, 24, 47, 101
maps *xi*, 1, 5–7, 15, 33, 45–6, 48, 52, 149–82
as inscription (something to be read) 33
Betts (mapmaker) 154
Colonial adventurers' maps 6–7, 152
mapping and the Victorian landscape 6, 149–82
maps of zoos 159–60
markings 32–4, 64, 74, 93–4, 106–07, 136, 152, 159, 167, 180
Marks, Laura U. 135
Marsden, Algernon Moses 130, 132
Marsyas 10
mascot 128
mastery 49–50, 163, 180
"Matabele War" panorama (1897) 23–5. *See also* Belle Vue
material *xi–xii*, 9, 31–2, 48, 50, 65, 101, 104, 143, 155, 159, 166–7
Matoppo (Matobo) Hills 24
Mayer, David 23
menageries 4–5, 18, 21, 23–4, 31–4, 40, 42–3, 45, 47, 49, 51, 56, 60, 62, 67, 75, 88–9, 93–5, 107–08, 111–13, 120–21, 128, 131, 134, 147, 178
Merleau-Ponty, Maurice 15, 129, 133
metamorphosis 130–31
Michelangelo 10
Last Judgment 10
Michie, Helena 135
Midgley, Mary 9, 130
mill 21, 29, 41, 47, 51
missionary 12, 49, 77, 155
Mitchell, D.W. 60
Mitchell, P. Chalmers 41
"Mons" panorama 24, 26. *See also* Belle Vue Zoo
Monster Globe Stand 22–3
monstrosity 20, 22–5, 46–7, 61, 160
Montagu, Ashley 15, 128
Morris, P.A. 40, 63, 79, 92
Murray, Henry 139
museums *xi–xii*, 4–5, 11, 17–18, 31, 40–49, 51–4, 57, 62–3, 65, 69, 71–9, 81–3, 85, 88–9, 93–8, 101–02, 104–09, 111–14, 116, 121, 133–4, 141, 159, 179, 181
and touch 17, 101, 121, 133–5
British Museum 49, 57, 72–3, 112, 134
Bullock Museum (Liverpool) 77

Horniman Museum (London) 106
Knowsley Menagerie Museum 88, 111, 113
Leverian Museum (London) 77
National Maritime Museum (Greenwich) *xi*, 78–9, 81, 83
National Museums Liverpool *xi–xii*, 53, 63, 69, 75–7, 79, 82–9, 98, 107, 113, 141
National Portrait Gallery 5, 18, 27–32, 45
Royal Menagerie (Windsor) 32, 67
Royal Ontario Museum (Toronto) 121
Smithsonian American Art Museum 65
South Kensington Museum 134
Walkley Museum (Sheffield) 109
Wombwell's Menagerie (Edinburgh, Scotland) 43
World Museum *xi*, 76, 78
Zoology Museum (Cambridge University) 101–02, 116

National Maritime Museum (Greenwich) *xi*, 78–9, 81, 83
National Museums Liverpool *xi–xii*, 53, 63, 69, 75–7, 79, 82–9, 98, 107, 113, 141
National Portrait Gallery 5, 18, 27–32, 45
natural history 3–6, 41–2, 45–6, 49–53, 56, 60–63, 65, 68, 73, 75, 78, 81–2, 85, 87–8, 92–4, 103–04, 107–14, 116, 155, 159–60, 163, 165, 181
natural history illustrations 5–6, 93, 107, 109, 111–14
Natural History Society (Manchester) 41
naturalism 41, 49–50, 52–3, 56–7, 61, 69, 71 82, 94, 104, 106, 109, 113, 130
New Zealand 75, 107, 161
Newcomen, Thomas 29
Nicholls, Robert 20–22, 29, 47
Nixon, Julie V. 165
nomenclature 12, 52–3, 64
nonsense verses, stories, and nonsense drawings 5–6, 41, 53–4, 87–120
North America (continent) 75, 80, 88, 129, 163, 171

O'Farrell, Mary Ann 95
Ocean Wave 22, 47. *See also* Belle Vue Zoo
Ojibbeway Indians 126, 131–2
order 11–12, 23, 50, 52–3, 67, 72, 74–5, 85, 106, 117, 119, 166, 170
ORLAN 130
Orwell, George 115
Oswell, W. Edward 155–6
Oswell, William Cotton 155–6
otherness 7, 12, 34, 160, 181
Owen, Richard 33, 57–8, 64, 94, 111, 155
Memoir on the Gorilla 57, 94
ownership 6, 41, 97, 106, 113, 119

panorama 18, 24–5, 27
Papua New Guinea 69
Patchett, Merle M. 96
pattern 33–4, 45, 74, 94, 153, 160, 167, 170–72, 177–8
Patterson, Mark 15–16, 135, 137
Peck, Robert McCracken 88, 110
Peel, Robert (Sir) 29
Perricone, Christopher 135
Peru 80
pests 42, 51, 70–73, 78, 81–2, 156
Phillips, Catherine 165, 170–71, 175–6, 179
phrenology 12
physical atlas 161
physiognomy 12, 40
Picturesque Guide through the Regent's Park (1829) 121, 126–7
Pile, Steve 180
Plato 167
plumage 69, 104, 109
plume 4
Poliquin, Rachel 96–7, 101–02
"The Port of Manchester" panorama (1894) 27. *See also* Belle Vue Zoo
portraiture *xii*, 1, 4–6, 17–48, 57, 62, 64–5, 74, 87–120, 130, 133, 135, 137–9, 143, 146, 180
portrait galleries 5, 18, 27–32, 42, 45
theories of 5, 17, 134
postcolonial 4, 128

Prince Albert 30
Puerto Rico 61
Punch 35–9, 53–5, 66–7, 72, 91, 102–03, 105, 112, 127. *See also* Sanbourne, Linley (cartoonist)

Queen Victoria 28, 31, 93, 109

racism 69
Raffles, Stamford (Sir) 60
Reader, Simon 171
Reeve, Lovell 52, 57, 82
refreshment room 21
reptiles 43, 63, 160–61, 163–4
  alligator 76, 160
  boa constrictor 42, 62, 161
  crocodile 95, 161
  cayman 64
  lizard 34, 41, 42
  turtle 53, 75, 160
Reynolds, James 163
Rhodes, Frank 22
ringworm 10–11
Ritvo, Harriet 23, 49, 52–3, 128–9
Robbins, Bruce 136
Robinson, Arthur H. 159–61, 163
Rodaway, Paul 177
Rosenberg, Harold 17, 31
Rossetti, Dante Gabriel 97
Rowley, John 101–03
Royal Menagerie (Windsor) 32, 67
Royal Ontario Museum (Toronto) 121
Rubens, Peter Paul 130, 132, 140, 143
rug 4, 47, 57, 62, 130, 139, 146, 154, 166, 171
Ruskin, John 109, 140, 171–2, 179
Ryan, James R. 155, 159

Saint Bartholomew 10. *See also* Michelangelo; d'Agrate, Marco
Sanbourne, Linley 35
Savery, Thomas 29
Schlossman, Beryl 146
Selby, Prideaux John 109
sensuous geography 150, 165, 175, 177–8, 180
Serres, Michel 15, 133–4, 137, 152
Shrigley, David 103
"The Siege of Ladysmith" panorama 28. *See also* Belle Vue Zoo; Boer War
skeleton 27, 41, 43–5, 47–8, 94, 96, 98, 100, 104
skin
  and contagion 27, 50, 70–73, 132
  and contamination 6, 67, 73
  and decay 49, 67, 70, 74, 102, 171
  and disease *xi*, 10–12, 41, 70, 110
  and ethnic identity 12, 64
  and landscape 6–7, 15, 18, 20, 46, 52–3, 88–9, 119, 133, 149–51, 153–5, 157, 159, 161, 163, 165–7, 169–71, 173–81
  and medical training 11–13, 69, 82, 110
  and race 9, 12, 33, 69, 107,
  and tattoos 12–13, 96, 102, 114
  as barrier 14, 40, 130
  as boundary 13, 34–5, 40, 45–6
  as canvas 15, 136–7, 139–40
  as clothing or garment 33, 64–5, 95, 143
  as evidence 10, 65
  as identity 1, 4–5, 9–10, 12–14, 16–18, 32–4, 45–6, 50, 64, 74, 94–6, 108, 116, 130
  as memory 9, 15, 43–4, 48, 65, 79, 97, 106, 115, 135, 140, 157
  as metonym for being 4, 10, 14, 33–4, 66, 74, 95, 136, 154
  as passport 13
  as protective envelope 4, 9, 13–14, 34–5, 39–40, 43–6, 49, 66–7, 81, 85, 97, 116, 130, 134, 140, 151, 157, 161, 176–7
  as site of encounter 1, 9, 12–14, 16, 31, 35, 73, 129, 132, 136, 146, 154, 176
  as trophy 4, 6–7, 23, 42, 45, 50–51, 56, 62, 65, 68, 71, 73, 76, 90–91, 149–50, 154–7, 166, 181
  as "worn identity" 4, 10, 14, 33, 64
  clinical attention to 10–11
  cognitive function of 4, 128, 133
  display of 1, 5–6, 9–10, 12, 14, 17, 31, 33–4, 42, 45, 49, 65, 67, 70, 103, 111, 114, 130, 140, 149,

in portraiture *xii*, 1, 4–6, 17–48, 57, 62, 64–5, 74, 87–120, 130, 133, 135, 137–9, 143, 146, 180
moral condition 10, 95
removal of/skinning 7, 10, 48–9, 60–63, 65, 67–70, 79, 82, 96, 98, 100–102, 138, 155–7. *See also* flaying
"second skin" 6, 15, 95, 130–31, 143, 146
surface 6–7, 9, 12, 14–15, 33–5, 43–5, 64, 74, 94, 98, 101, 106–08, 110–11, 113–14, 131, 134, 136, 139–40, 142, 152, 156, 161, 167–8, 177, 180
theories of 9–16, 32–3, 35, 45, 64, 66, 128, 132, 134, 136
wax replicas of *xi*, 11
Smith, Adam 77
Smithsonian American Art Museum 65
South Africa 65, 73, 75, 77, 96, 121, 155
South America (continent) 5, 32, 61, 75, 78–80, 82, 87, 89, 103, 160–61, 163
South Kensington Museum 134
souvenir 5, 45, 62, 65
species differentiation 9, 32, 34, 64
specimen 1, 3–5, 9, 32–3, 40–43, 46–7, 49–53, 56–7, 60–65, 67, 69–79, 81–2, 84–5, 88–9, 93–4, 97–8, 100–107, 109–14, 120, 159–60, 163, 181
spectacle 23, 25, 27–30, 46, 64, 104, 111–12, 114–15
Spencer, William 143, 161
Sportsman's guides or handbooks 56, 70, 91, 155, 157
St. John's Hospital for Diseases of the Skin (London) 11
Stacey, Jackie 14, 33, 35, 128–9, 134, 136
Stanhope (Lord) 30
Stanley, Edward Smith (13th Earl of Derby) *xi–xii*, 4–5, 49–50, 53, 57, 74–5, 77–9, 82, 87–9, 93, 108, 113, 117
Stephenson, George 29
Stevenson, Robert Louis *ii*, 40, 56
The Strange Case of Dr Jekyll and Mr Hyde 40
Stewart, Susan 62, 66
Stoichita, Victor L. 136
Stradanus, Giovanni 10
Apollo Flays Marsyas (ca. 1580–1600) 10
Sumatra 60
superiority 23, 35, 52, 68, 113, 116
surface 6–7, 9, 12, 14–15, 33–5, 43–5, 64, 74, 94, 98, 101, 106–07, 110–11, 113–14, 128, 131, 134, 136, 139–40, 132, 152, 156, 161, 166–7, 172, 175–7, 179–80
Surinam 61
surveillance 140, 143
Sussman, Herbert 13, 20
Swainson, William 33, 63, 68, 72–3, 94–5, 100
Swayne, H.G.C. (Major) 57, 156
Swift, Jonathan 149, 154
syphilis 10–11

Tallis, Raymond 15, 132–3, 145, 147
tattoos 12–13, 96, 102, 114
Macdonald's tattoo studio (Jermyn Street, London) 13
taxidermy 14, 40–42, 47, 56, 63, 67–8, 72–3, 79, 90–94, 96–8, 100–105, 110, 114, 120, 156
manuals of 5, 63, 79, 90, 92, 98, 100
taxonomy 18, 48, 53, 77
technology 14, 17–18, 20–21, 27, 41, 52, 100, 129
Tennyson, Alfred 136
textile 47, 65, 130
texture 6–7, 12, 15, 32–3, 43–5, 65, 93–4, 96, 106, 109, 120, 129, 131–3, 135–6, 139, 143, 145, 153, 157, 166, 169, 177–80
Thames (river) 10, 166
The Ladies Manual of Art (1890) 79
The Pictorial Museum of Animated Nature (1843–1845) 45, 121
The Strand Magazine: An Illustrated Monthly 13
theater 5, 173

Thompson, E.P. 51
Thrift, Nigel 180
"Tippo's Tiger" 95
Tissot, James 130
Titian 10, 142–3
  Flaying of Marsyas (ca. 1570–1576) 10
tongue 6, 78, 104, 121, 167, 176
touch *xii*, 5–7, 13, 15–17, 44, 46, 72–3, 96, 101, 121, 123, 125–40, 142–3, 145–7, 150, 152, 156, 170, 175, 177–81
  theories of 15, 128, 132, 134
Towne, Joseph xi
trade unions 21
transportation 17, 47, 60, 62
  of exotic animals 4, 7, 60, 62–3, 65, 75, 89, 126, 155
travel narratives 5, 69, 161. *See also* Hornby, Elizabeth
travelers 5, 13, 51, 56, 70, 126
treatise (scientific) 5, 11, 139
trophy 4, 6–7, 23, 42, 45, 50–51, 56, 62, 65, 68, 71, 73, 76, 90–91, 149–50, 154–7, 166, 181

unionists 21

velocipede 22, 47
Victorian
  agents 63, 70, 75–6, 77, 82, 84–5
  hunters 5–6, 11, 34, 51, 56–7, 63–5, 71, 73, 83, 90, 96, 98, 106, 116, 149, 152–7, 159, 161, 180–81
  maps *xi*, 1, 5–7, 15, 33, 45–6, 48, 52, 149–82
viscera 94, 96
vision 133, 166, 171
von Uffenbach, Zacharias Conrad 134
vulnerability 4, 9, 34–5, 46, 74, 78, 85, 94, 101, 140, 151

Wade, H.T. (map publisher) 156
Walkley Museum (Sheffield) 109
Wallace, Alfred 49, 52, 60, 67–9, 160, 163
Wallace, Ian 77
Ward, Priscilla 70
Ward, Rowland 56, 68, 70, 91–4, 156–7, 166
  Wardian furniture 91–4
Waterton, Charles 61, 63, 104–05
Watt, James 29
Wellcome Institute (London) 96, 102, 114
West Africa 57, 63
Western Dispensary for Diseases of the Skin (Fitzroy Square, London) 11
Whatmore, Sarah 50, 180
Whistler, James McNeill 5–6, 137–9, 142–6
  and Joanna Hiffernan 142–3
  Symphony in White, No 1: The White Girl (1862) 6, 137, 142, 144
Whitaker, Katie 51, 75
Whitfield, Thomas 76
Wilde, Oscar 51
Wills, John Walpole 77
Wilson, Erasmus 11
  A Practical Treatise on Healthy Skin: with Rules for the Medical and Domestic Treatment of Cutaneous Diseases (1844) 11
Wolf, Joseph 87, 93, 96, 108–09, 112
Wollstonecraft, Mary 176
Wombwell's Menagerie (Edinburgh, Scotland) 43
women 11, 21, 51, 78–80, 106, 132, 142, 151, 173
Wood, Samuel 68
Woodall, Joanna 30, 109
World Museum *xi*, 76, 78

Zaniello, Tom 165–6
Zimmern, Helen 139
zoogeography 6, 149–50, 159–61, 165–7, 170, 175, 179, 181
Zoological Society (London) 13, 18, 41, 45, 49, 51, 57, 60–61, 63, 65, 68, 72, 75–6, 78, 87, 91, 96, 109, 114, 126–7, 131, 159
Zoology Museum (Cambridge University) 101–02, 116
zoo *xi–xii*, 1, 4–6, 13, 17–29, 31–43, 45–9, 52, 54, 61–3, 74, 76–7, 93–4, 96–7, 107, 109, 120–23, 125–30, 132–3, 145, 147, 159, 178–81
  Belle Vue Zoo (Manchester) *xi–xii*, 5, 17–35, 37–48, 62, 96, 127, 149, 180–81

Berlin Zoo 63
Bristol Zoo 18
Buffalo Zoo 62
London Zoological Society Gardens 13, 18, 41, 45, 49, 51, 57, 60–61, 65, 68, 72, 75–6, 78, 87, 91, 96, 109, 114, 126–7, 131, 159